THE NEW NATURALIST

A SURVEY OF BRITISH NATURAL HISTORY

THE WORLD OF THE HONEYBEE

The aim of this series is to interest the general reader in the wild life of Britain by recapturing the inquiring spirit of the old naturalists. The Editors believe that the natural pride of the British public in the native fauna and flora, to which must be added concern for their conservation, is best fostered by maintaining a high standard of accuracy combined with clarity of exposition in presenting the results of modern scientific research.

Plate I.—A Honeybee Community (*Apis mellifera*)
The multi-comb nests of the Western Honey bee are normally built in hollow
trees and similar sheltered places although occasionally a swarm will build a
nest in the open air, presumably because it fails to find a suitable protected site.

THE NEW NATURALIST

THE WORLD OF THE
HONEYBEE

by

COLIN G. BUTLER
O.B.E., PH.D., F.R.S.

WITH 2 COLOUR PHOTOGRAPHS
AND 87 BLACK AND WHITE PHOTOGRAPHS
TAKEN BY THE AUTHOR

COLLINS
ST JAMES'S PLACE, LONDON

First edition 1954
Reprinted 1958
Revised edition 1962
Reprinted 1967
Reprinted 1971
Revised edition 1974
Reprinted 1976
Reprinted 1977

ISBN 0 00 219552 6
Printed in Great Britain by
Collins Clear-Type Press, London and Glasgow

"Inter omnia insecta principatus apibus, et jure praecipua admiratio."

PLINY. *Lib. II. c.5*

CONTENTS

LIST OF PLATES

ix

PREFACE TO THE THIRD EDITION

It is almost exactly twenty years since the first edition of *The World of the Honeybee* was published—twenty years during which many exciting advances in our knowledge of the behaviour and physiology of the honeybee have been made. These include much more information about the chemical compounds (pheromones) produced by larvae as well as adult queens and workers and of their role in controlling the behaviour of the individuals of which a honeybee colony is composed, and thus of the colony as a whole. The sex attractant released by the queen which enables flying drones to find her has been discovered and found to be the same substance as the principal component of the "queen substance" which I postulated in 1954. "Queen substance" in its turn is now known to be a mixture of several substances most, but perhaps not all, of which are found in the secretion of the queen's mandibular glands; substances that not only inhibit unnecessary queen rearing by the worker bees and development of their ovaries, but also help the workers of swarms to find their queens and cluster quietly with them.

The truth of the old idea that the worker bees of a colony are able to distinguish their own queen from similar queens belonging to other colonies has at last been demonstrated experimentally, as has the much greater part played in swarming and foraging, as well as home-finding, by the scent produced in the Nassanoff glands of worker bees than had hitherto been supposed. The composition of the Nassanoff scent, one of the first pheromones to be discovered, has also been elucidated.

The question as to how a queen honeybee distinguishes between worker and drone cells in the darkness of the hive and so is enabled to lay fertilised and unfertilised eggs in them, respectively, has also been discovered.

We now know more about the ways in which worker bees communicate with each other and how they recognise their own hives and colony members; and the part that substances that bees deposit inadvertently when running around their hive entrances—"footprint" substances—play in helping homecoming foragers to find the entrances of their hives has been recognised.

More has been learned about direction finding and foraging, as well as about caste determination. Indeed, it seems as though we must at last be on the verge of discovering the secret of the way in which the queen and worker castes are determined.

Although few of the new discoveries that have been made by research workers in many different countries appear to have any immediate, direct application to problems of practical beekeeping, they are important because they enable beekeepers to recognise and discard many practices that have been based on ignorance and fallacy, and to adopt a much more rational approach to beekeeping than has hitherto been possible.

I am very grateful to the Editors of the *New Naturalist* series and to my Publishers for allowing me to make so many changes in the text and, indeed, to rewrite many parts of it, in order to bring it more nearly up-to-date, although even small changes in the text are very expensive to make and must necessarily increase the price at which a book can be sold. Because of the expense that would have been involved I have not made several minor alterations which, although I think them desirable, are not essential. I have not, for example, altered the specific name of the Eastern honeybee from *Apis indica* to *Apis cerana* in order to conform with the rules of taxonomic nomenclature, nor have I changed "queen substance" to "queen substances".

Finally, I hope that I may be allowed to echo the words of Gilbert White who wrote: "These observations are, I trust, true in the whole, though I do not pretend to show that they are perfectly void of mistake, or that a more nice observer might not make additions, since subjects of this kind are inexhaustable."

Harpenden, 1974.   C.G.B.

CHAPTER I

THE ORIGIN OF THE HONEYBEE
AND EVOLUTION OF THE RELATIONSHIP
BETWEEN BEES AND FLOWERS

THE HONEYBEE is a social insect, that is to say an insect which lives in a society. It is, therefore, desirable that we should be quite clear what we mean by an insect society, before we begin to consider how such a complicated organisation as a honeybee society came into being. Unfortunately it is very difficult to frame a comprehensive and exact definition of social life. Indeed, at a meeting of the British Association some time ago at which the behaviour of social insects was discussed, nobody was able to put forward a definition which was acceptable to everyone present and which was considered to be entirely adequate.

Some people may say, "Well, why worry so much about trying to define exactly what you mean by social life; we know that the honeybee is a social insect, that the ants are social insects, and that at least some of the wasps also lead social lives: and surely provided that one knows these things there is no need to worry about a definition?" Now, for many purposes we can get along quite well without any definition, knowing that a honeybee is called a social insect without understanding why this is so. However, in order that we may be able to distinguish those insects which live social lives from those which do not, it is desirable that we should have a definition. Probably the most accurate and comprehensive definitions of insect social life are those that have been put forward by Dr. W. Morton Wheeler and which have been strongly supported by Dr. A. D. Imms. One of these definitions states that "a social insect is an insect which lives in a society;

each society consisting of the two parents (or at least the fecundated female) and their offspring; the two generations living together in a common abode or shelter and exhibiting at least some degree of mutual co-operation." If we accept this definition, as I think we must, it follows that amongst the social insects there is necessarily a lengthening of the lives of the adults, at any rate of the colony mother, since she at least lives sufficiently long to associate with her offspring; whereas, in the great majority of insects the mother dies without having had the least contact with her offspring, indeed in many cases before they have even hatched from their eggs. Such lengthening of the life of the colony mother has undoubtedly been associated with a degree of specialisation for a particular type of work at the expense of other types. The degree to which such specialisation has taken place varies very greatly, however, amongst the different groups of social insects— in some cases even within a single group. In the case of the queen honeybee specialisation for egg production has gone a very long way. This has not been brought about by acquisition of the ability to undertake a new type of work, but rather by the gradual suppression of any tendency to perform the other types of work that were previously carried out by her more primitive ancestors. The primitive ancestors of the queen honeybee undoubtedly undertook such duties as nest-building, foraging for food and care of the young; duties which she no longer performs.

Most social insects of the present day—the ants, bees, wasps, and termites or so-called white ants—build their nests in such dark and ill-ventilated places as (for example) hollows in trees, crevices in walls and rocks, and holes in the ground. Wheeler has suggested that these situations, and the very nature of many of the nests, may exert some influence towards extending the length of life of the individuals. It would perhaps be interesting to compare the length of life of workers and queens of *Apis dorsata*, a

---

Plate 1 (opposite). NEST OF A "STINGLESS" BEE (*Melipona iridipennis*)
Colonies of this tiny social bee live in Ceylon and India and build their nests in holes in walls, letter-boxes, etc. This nest is in an empty bee-hive and is made from a mixture of wax and plant resins. No combs are built, but the brood cells x and the food storage cells y are quite distinct. (x 1·0) (see also Plate 2).

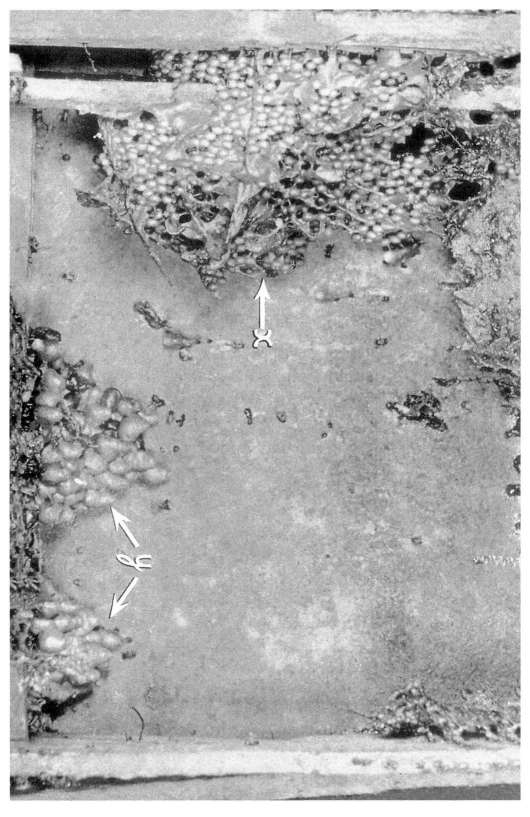

Plate 2

species of honeybee which builds its single comb nest in well-ventilated places in the open air, with those of workers and queens of *Apis indica*, a species which lives in the same district, but normally nests in cavities in trees and is much more highly specialised.

We must now repeat the definition of a social insect in slightly different terms. *All true insect societies are families:* that is to say (to use Dr. Wheeler's words) "affiliations of parent or parents, in most cases of the mother alone, with their offspring." Such a definition leads to the conclusion that insect social life in any highly developed state, such as that shown by a colony of honeybees, must have evolved slowly through numerous stages, showing an increasing degree of intimacy between the mother insect and her offspring, and with it an ever-increasing degree of mutual co-operation. Thus any idea, such as that put forward by the great naturalist Fabre and others, that the modern societies of ants, bees and wasps may have arisen from chance associations of female insects of the same species seems very unlikely.

Now although palaeontology (the study of the fossilised remains of plants and animals), can teach us a great deal about the evolutionary history of insects, including that of the social insects, it clearly cannot tell us very much about the evolution of social life itself. Behaviour cannot be preserved in fossil form. Therefore, in our attempts to discern the probable stages in the evolution of social life, as exemplified in the complex organisation of the honeybee colony, we must study the social life of other, less specialised, bees that are still living today.

Before we consider the evolution of social life amongst insects any further, let us consider the origin of the honeybee herself and her relationship with other insects.

The European honeybee, *Apis mellifera*, belongs to the great Order Hymenoptera, of which it has been estimated that some 100,000 species exist in the world today. This Order can for

---

Plate 2 (opposite). DETAILS OF THE NEST OF A "STINGLESS" BEE (*Trigona iridipennis*)

a (top). Workers guarding mouth of entrance tunnel of nest. (x 2.5)

b (centre). Group of brood cells. (x 2.5)

(lower). Group of food storage cells. (x 2.5).

convenience be subdivided fairly naturally into two Sub-Orders, the Symphyta and the Apocrita. The first of these Sub-Orders, the Symphyta, consists of the Sawflies and their allies; hymenopterous insects which possess no constriction or waist between thorax and abdomen, and of which the ovipositor has become modified to form a saw with which the female insect cuts slits in the tissues of plants to serve as nurseries for her eggs.

The members of the second Sub-Order, the Apocrita, all have a constriction or waist separating thorax from abdomen. The Apocrita in their turn can usefully be subdivided into two large groups, the Parasitica, so-called because their larvae live as parasites on other insects, and the Aculeata or stinging hymenoptera, to which group the Ants, Bees and Wasps belong.

All authorities appear to be agreed that the Symphyta both in structure and habits are more primitive and more nearly ancestral than the Aculeata, the members of which together form the most recent and most highly specialised group of the Hymenoptera.

Now let us consider some of the fossil insects in an attempt to decide upon the origin of the Hymenoptera as a whole. In dating the Geological Column I follow Dr. L. D. Stamp's *Britain's Structure and Scenery* in this series.

The oldest fossil insects known are about three hundred million years old. Fragments discovered in flakes of Rhynie Chert from the middle Old Red Sandstone of Scotland (Mid Devonian) appear to belong to the Order Apterygota, primitive wingless insects of which the well-known, abundant, and widely distributed Springtails are examples. However, the abundance and in some cases degree of specialisation of those Pterygota, or winged insects, which have been found in the Upper Carboniferous rocks has, according to Dr. A. D. Imms, led palaeontologists to conclude that these winged insects must have been developed from wingless ancestors not later than Devonian or Lower Carboniferous times, about 280 million years ago.

For many years palaeontologists were unable to find any primitive fossil insects from which it appeared to be at all probable that our modern Hymenoptera could have been derived. At last, however, in 1924 Dr. Tillyard described the remains of some fossil insects which had been discovered in Permian rocks (about two hundred million years old) in Kansas. These were mainly beautifully preserved wings. Tillyard called this newly discovered Order of fossil insects the "Protohymenoptera" and came to the conclusion that they were probably ancestral forms from which our modern Hymenoptera were eventually derived.

The most primitive of the three species of "Protohymenoptera" found at this time had wings which were in some ways like those of some of our modern Sawflies. In all these fossil forms the two pairs of wings were much alike in size and shape, and were stiff and glassy like a honeybee's wings, but, unlike them, did not possess any mechanism for coupling together the two wings of a side. Tillyard thought that some simple evolutionary changes probably led from these ancient forms to the modern Sawflies and their allies. The first true Hymenoptera to be discovered were found in rocks of the Jurassic period (150 million years old or less) and were, Tillyard thought, derived from the "Protohymenoptera" of the Lower Permian, and perhaps Upper Carboniferous, times. The Aculeata (or stinging) Hymenoptera, to which group the honeybee belongs, was one of the last great groups of insects to be evolved. However, by the Tertiary Era both parasitic (Parasitica) and stinging (Aculeata) Hymenoptera had become abundant.

More recent workers have maintained that Tillyard's conclusions are incorrect and that his "Protohymenoptera" were misnamed and cannot be the ancestors of modern Bees.

Dr. Imms has pointed out that the structure and behaviour of some of the more primitive families of wasps suggests that it is amongst them that we should look for the ancestors of both the bees and the ants as we know them today; and that, as a group, bees are probably best regarded as wasps that have forsaken a carnivorous diet in favour of a vegetarian one and, in the process, have become beautifully adapted for the collection of nectar and pollen.

In certain Cretaceous rocks many fossil plants including the remains of flowering plants are found, and since the evolution of insects must necessarily have followed that of plants, and the evolution of the nectar and pollen feeding bees that of the flowering plants, it is probable that the bees were developed from some wasp-like ancestor somewhere about the time that the Upper Chalk was deposited, perhaps eighty million years ago. It was among rocks of the Miocene period (33 to 13 million years ago) that Dr. Heer in 1849 discovered the fossil bee which he called *Apis adamitica*, which is one of the oldest bees known to science. Unfortunately, however, Heer was incorrect in placing this little bee in the genus *Apis*, the genus to which all species of honeybees living today belong. According to that great American authority on bees, Dr. T. D. A. Cockerell, Heer's bee really belongs to the genus *Lithurgus*, a genus of bees related to our familiar leaf-cutting bees. Similarly von Buttel-Reepen's belief that some of the fossil bees found in Baltic Amber are honeybees has also been disproved by the researches of Dr. Cockerell.

We must conclude regretfully that there are still many missing steps in our knowledge of the origin of our modern honeybees.

Evolution of the relationship between bees and flowers.

We have seen that the Bees (well-named Anthophila or flower-lovers) probably evolved from some wasp-like ancestor shortly after the first of the flowering plants began to appear. It is indeed more than likely that the flowers and the bees evolved more or less contemporaneously, and it is certain that the Bees have had a great deal to do with the evolution of the flowers as we know them today. This is an aspect of the bee-flower relationship which has frequently been overlooked and it is, therefore, perhaps worth while to examine it briefly.

Botanists tell us that some of the earliest flowers were almost certainly very similar in form and structure to the flowers of such plants as the magnolias and the buttercups as we know them today. They were flowers which at first probably had no petals, but certainly possessed numerous stamens and produced very large quantities of pollen. Petals, which may or may not have been coloured at first, were possibly developed by modification of some of the outermost stamens. They probably served to prevent the precious pollen from getting unduly wet and thus into an unsuitable condition for transference to another flower.

Now it is a well-established fact that cross-pollination, which is, broadly speaking, the transference of pollen from the flower of one plant to a flower of another plant of the same species, has always been highly advantageous to the progress of the species. The development by a plant of some new and more efficient method which leads to a greater degree of cross-pollination than hitherto, in contrast to pollination with pollen from its own flowers, is one of the most valuable weapons in the eternal struggle for existence. Provided, as seems probable, that the primitive bees were able to distinguish between different colours, one can imagine that when a few plants of a group whose flowers were habitually visited by bees developed conspicuously coloured petals or stamens the bees were especially attracted by these flowers, although by no means exclusively as there would be few such flowers at first. Such exceptional plants had, by chance, found a way of encouraging the cross-pollination of the flowers and, therefore, achieved a greater degree of success in the struggle for existence than the rest

We have been assuming that the primitive bees were in the habit of visiting the primitive flowers. We must consider now how and why this habit began. Perhaps the pioneer bees were in search of shelter from rain or dew; but it is possible that they were in search of nectar, a habit that they had inherited from their wasp-like ancestors. Many of our modern wasps, especially the social species, visit flowers regularly in search of nectar. (Pl. 5, p. 38). Carbohydrates (nectar, honeydew, honey, jam, etc.) are not mere *bonnes bouches* to present-day social wasps; they form a necessary part of their diet, and are collected and carried home to the nest with great perseverance and regularity.

It is easy to imagine how the primitive wasps are likely to have acquired the habit of collecting nectar. They were fundamentally carnivorous, as almost all of them are still. Perhaps when searching for insects they stumbled upon some honey-dew, the sugary liquid excreted by greenflies and their allies. Perhaps they found drops of nectar on the extra-floral nectaries of some plant. Being thirsty, they drank some of the syrupy liquid, and found it to their liking.

When visiting flowers in search of nectar the primitive bees inevitably got their bodies dusted with some of the abundant pollen that was present. It does not require any great stretch of the imagination to visualise how they acquired a taste for the pollen, which is very nutritious food with a high protein content, possibly when cleaning it from their bodies with their mouthparts.

The primitive bees probably began, in some such way, to feed both on pollen and nectar to the exclusion of other kinds of food. Since flowers were still a relatively unexploited source of food, they found a niche for themselves and became very successful in the struggle for existence.

As the bees became more and more dependent upon flowers for their source of food, those of them with slightly longer tongues, or slightly more hairy bodies, were more successful than their shorter tongued and less hairy contemporaries. Structures for the collection of nectar and pollen were gradually developed and finally culminated in the highly specialised organs of the worker honeybee as we know her today. However, by no means all of the bees developed such highly complex structures as those of the honeybees, and we still find existing today primitive little bees of the family *Prosopidae*, which have mouth-parts very similar to those of the wasps and which possess no specialised organs for the collection of pollen. The *Prosopidae* have to swallow the pollen they collect, together with the nectar, in order to transport it home to their nests.

How early nectar came to serve as an attraction for insect visitors we have no way of knowing. Probably the primitive bees visited all the species of flowers available to them quite indiscriminately, especially as they were all much alike. Gradually, however, natural selection favoured the development of differences between flowers—differences of colour, form, odour, and season of flowering (this last is most important). As the forms of the flowers became more varied so the flower-visiting insects became more specialised for the collection of food from particular types of flowers and thus less able to collect food from other types. Some insects learned to associate particular colours and perfumes with the presence of food and to neglect other perfumes and colours until, eventually, as in the case of certain bees today, special perfumes and colours became almost exclusively attractive to them. The flowers of some plants and the bodies of some kinds of insects became so closely adapted to suit one another that these flowers and insects eventually became completely dependent upon each other. The flowering times of some species of plants and the times of emergence of the adults of some flower-visiting insects also became very closely suited. On the whole, however, the bees did not become so closely tied as some other insects have done to specific flowers, and continued to seek food from a wider range of species. It is certainly advantageous to plants that insects should become so highly adapted to their specialised flowers as to be caused to concentrate upon them to the exclusion of other species.

Now each worker honeybee, although she may visit a number of different kinds of flowers in the course of her foraging life, usually, but not quite always (p. 193) confines her attention to a single species on any one foraging expedition. At first sight it might appear that a bee which visits more than one species indiscriminately on any one foraging expedition, would be a much less efficient pollinating agent than another bee which remained constant in her attentions to one particular species, even though she may at times assist the process of evolution and the development of new species of plants by causing hybridisation. However, as Miss A. D. Betts has pointed out, a bee which visits the flowers of two species of plants (let us call them A & B) in the course of a single foraging expedition, will usually pollinate both of them successfully. Both A and B are likely to be pollinated almost as effectively as if the bee had been concentrating her attention on only one of these species at a time. The flowers of the plants of species A will receive sufficient pollen from other specimens of their own species to ensure that they are pollinated, and the flowers of species B will likewise receive sufficient of their own kind of pollen. Even should these two species of plants, A & B, be sufficiently closely related to one another to render hybridisation possible and should a flower of species B receive, more or less simultaneously, pollen from flowers of both species A & B, hybridisation will, nevertheless, only occur very infrequently, as the pollen grains from other flowers of species B will almost certainly be more potent than pollen grains from species A and their tubes will reach and fertilise the ovules before those of A can do so. Thus even should a bee visit two species of flowers quite indiscriminately she will not, as has sometimes been suggested, be useless as a pollinating agent; she will on the contrary be quite reasonably effective. Admittedly pollen will be wasted, but in this connection we must remember that many species of plants "waste" pollen, in the sense that they produce quantities far beyond their reproductive needs. Most of it is "bait"—an attraction to the necessary pollinating insects which carry it away to their nests in large quantities for food, and only transfer a little of it to the stigmas of other flowers. Some plants, such as the toad-flax, have developed mechanisms which greatly reduce the amount of pollen which nectar-seeking bees can "waste". In the case of toad-flax flowers the stamens are placed in such a position that the nectar-seeking bee receives the pollen on a part of her back that she has great difficulty in cleaning, a position in which the pollen grains come into contact with the stigmas of further toad-flax flowers directly she enters them in search of nectar. Other flowers such as those of the lime-tree produce sufficient nectar to attract bees who pollinate them whilst collecting it, but for some reason the bees seldom collect pollen from them.

It is interesting to contemplate the fact that many of our flowers in their wonderful variety, with their many forms, scents and colours, owe their very existence to the evolution of bees.

CHAPTER 2

THE HONEYBEES
AND THEIR DISTRIBUTION TODAY

APPRECIABLE QUANTITIES of honey and wax can be obtained only from colonies of bees belonging to the single genus *Apis* and commonly known as honeybees. There are other kinds of bees which collect and store honey in such a way that man can obtain it from their nests, such as the *Meliponinae* or so-called "stingless bees" (Pl. 1, p. 2 and Pl. 2, p. 3) and the bumblebees (*Bombidae*) (Pl. 3, p. 18), but the quantities of honey which they store are so small that these bees are almost useless to man as honey producers. Furthermore their honey is often of very poor quality and is stored in their nests in such a way that it is difficult to extract it hygienically.

Only four species of the genus *Apis* (honeybees) are found in the world today. Three of these four species occur abundantly in the wild state both in the jungles and cultivated areas of Southern India, Ceylon and other parts of South Asia, which is, undoubtedly, their natural home. These are the Giant honeybee or Bombara (*Apis dorsata*), the Little honeybee (*Apis florea*) and the Eastern honeybee (*Apis indica*).

The fourth member of the genus *Apis*, the Western honeybee (*Apis mellifera*) does not occur wild in any part of South Asia but is found in the wild state in Europe and possibly also in North Africa. It should be realised, however, that all our honeybees are "wild"; man has never succeeded in domesticating the honeybee in the way in which he has domesticated some other animals. All that man has been able to do has been to induce colonies of two of our honeybees, *Apis mellifera* and *Apis indica,* to build their nests in conveniently accessible places by providing them with suitable types of shelters, such as modern hives, so that he can readily rob them of their stores of honey. The colonies of the other two kinds of honeybees, the Giant honeybee and the Little honeybee, which build single-comb nests in the open air, have never been induced to live happily in hives. It is only because *A. mellifera* and *A. indica* prefer to occupy cavities in trees and similar places, and build their multi-comb nests in them, that man has succeeded in inducing them to live in his hives. A colony of honeybees living in the most up-to-date type of hive is no tamer, no more "domesticated", than its sister colonies inhabiting hollow trees in nearby woods and jungles.

The Western honeybee (*A. mellifera*) and the Eastern honeybee (*A. indica*) (Pl. 11, p. 82) are certainly very closely related to one another. Because of anatomical differences, however, they are certainly better regarded as distinct species, rather than sub-species of the genus *Apis*. Their behaviour is in most respects very similar, and workers of *A. mellifera* have been known to look after an *A. indica* queen. It seems likely that *A. indica* and *A. mellifera* were derived from a common ancestor somewhere in South Asia in geologically recent times, and that, of the two, *A. indica* is slightly nearer to this ancestral type of bee in its behaviour.

As the colonies of this ancestral type of honeybee spread outwards from their ancestral home in South Asia, they became adapted to the different climatic and floral conditions which they found in the new areas which they colonised. The Western honeybee (*A. mellifera*), as found in Britain and elsewhere in Europe today, gradually modified certain of the behaviour patterns which she had inherited from the ancestral honeybee, and partially or completely suppressed others. On the other hand the Eastern honeybee (*A. indica*) in the course of her evolution tended to depart less from the ancestral type, since she continued to live in a tropical environment similar to that inhabited by her ancestors for thousands of years.

The results of these evolutionary trends are that the Eastern honeybee (*A. indica*) still retains a number of ancestral behaviour patterns which are of great value and importance for her con-tinued survival under the environmental conditions in which her colonies exist at the present time. On the other hand the Western honeybee (*A. mellifera*) has some of these behaviour patterns almost entirely suppressed or modified; this is of the greatest significance to the survival of her colonies in the harsher climatic conditions of the temperate and sub-arctic areas of the world in which she lives almost exclusively today.

A subspecies of *Apis mellifera*, known as *adansonii*, is found widely distributed in Africa. So far very little seems to be known about its behaviour; but I suspect that it will be found to be somewhat intermediate between that of *Apis indica* and the typical Western honeybee.

A good example of a behaviour pattern which is still retained by *Apis indica*, and which plays an important part in the survival of her colonies in Ceylon and India, is the habit of the *whole* colony deserting a nest-site for a new one, or even for a new district altogether, when local conditions either within the nest or in the locality become unfavourable. Unfavourable nest conditions can be caused by destruction of the combs by wax-moths (Pl. 4, p. 19) or by invasion of the nest by termites or ants. A temporary lack of nectar-producing flowers also renders a locality unfavourable to colonies which have not stored away an adequate reserve supply of food to meet their requirements during the period of shortage. This "absconding" behaviour also assists the survival of the two more primitive species of honeybees, *Apis dorsata* and *Apis florea*, which occur only in the tropics of South Asia.

"Absconding", in which all the adult members of the colony forsake one nesting place in favour of another, must not be confused with the method of colony increase known as swarming in which only a proportion of the adult members of the colony leave the nest to found a new one elsewhere (p.154.)

"Absconding" has become, almost, but not quite entirely, suppressed in the case of the Western honeybee. It would be very detrimental to colonies of this honeybee if absconding still persisted, as they would often perish if conditions were inclement. The remains of this behaviour pattern are still shown occasionally in the form of so-called "hunger swarms" and (perhaps) in those "mating swarms" in which all the worker bees accompany their young queen on her nuptial flight and, instead of returning to their hive, go off with her to build a new nest elsewhere; but such behaviour is rare, and a colony of Western honeybees, unlike their Eastern counterparts, will usually die of starvation in their own hive rather than leave it to seek their fortune elsewhere.

The other two honeybees, the Giant honeybee (*A. dorsata*) and the Little honeybee (*A. florea*) are quite distinct species from *A. indica* and *A. mellifera,* and are altogether more primitive and much less advanced in their behaviour, although they do have some similar behaviour patterns.

Both *A. dorsata* and *A. florea* always build their single-comb nests in the open air, never in the hollow tree or similar cavity which *A. indica* and *A. mellifera* use in ninety-nine cases out of a hundred under similar natural conditions. But it is interesting to note that *A. mellifera* will, very occasionally, build her multi-comb nest in the open air (Pl. 1, frontispiece). However, those colonies which revert to this primitive habit (which is obviously not completely suppressed) do not persist for very long in temperate climates.

All the kinds of honeybees that are found in Ceylon and other parts of South Asia exhibit a remarkable behaviour pattern when a hornet, ant, wax-moth (Pl. 4, p. 19) or other intruder approaches their nests. They shake their bodies violently from side to side in concert, a behaviour pattern which I have called the "shimmering" behaviour and which usually results in hasty retreat of the invader. The Western honeybee, which does not have to face these pests throughout the greater part of her range, no longer exhibits this behaviour pattern and, furthermore, is much slower and more deliberate in her movements when visiting flowers than the Eastern honeybees; she thus falls an easy prey to foraging hornets when introduced into Eastern countries. Whether those specimens of *A. mellifera* which live in Africa and the Middle East, where the hornet menace also exists, still retain this protective habit does not seem to have been recorded, but I strongly suspect that they do. It has also been suggested to me that those colonies of *A. mellifera* which live in Cyprus, and which are known as Cyprian bees as they constitute a distinct race, still retain this "shimmering" habit and certainly I have heard colonies of this race making the peculiar hissing sound which accompanies the "shimmering" behaviour of *A. indica*.

The Giant honeybee (*A. dorsata*) is much the largest of the honeybees, and the Little honeybee (*A. florea*) (Pl. 10, p. 67) is very much the smallest. The Eastern honeybee (*A. indica*), the hive-bee of the East, occupies an intermediate position and is decidedly smaller (as much as a third) than bees of the present-day races of the Western honeybee (*A. mellifera*) (Pl. 11, p. 82).

The colonies of the Giant honeybee build single, very large, combs attached to the undersides of overhanging rocks (Pl. 9a, p. 66), suspended from the more or less horizontal branches of tall trees, or sometimes even from the eaves of tall buildings. A single comb may be as much as 5-6 feet long and 3 feet deep. Honey is stored, sometimes in fair quantities (30-40 lbs.), in the upper portion of the comb, which may be as much as 4 inches thick. The brood is confined to the lower part of the comb, which is about 1¼ inches thick. The worker bees are reared in cells which are almost identical in size in cross section with the worker cells of the combs of the Western honeybee but which are nearly twice as deep. The worker of the Giant bee is almost twice as long as the worker of the Western bee and has dusky coloured wings (Pl. 9b, p. 66). Whereas in the case of the Western honeybee the drones, or male bees, are considerably larger than the workers (Pl. 6, p. 39) and are reared in cells of greater size, the drones of the Giant bee are about the same size as their sister workers and are reared in cells with similar dimensions.

Workers of the Giant bee are reputed to be easily disturbed and liable to be very ferocious and to attack man with little provocation and great persistence. There are, undoubtedly, a number of authentic cases on record of men and women being stung to death by groups of these bees despite the fact that (in the author's limited experience) the sting is little more painful, and the venom no more severe in its effect, than that of the Western honeybee. It seems, however, that the bees have to be considerably provoked before they will attack in this way, and they are readily subdued by smoke and can be handled safely if care is taken. It is also likely that some of the tragedies for which the Giant bee has been blamed should really have been attributed to an oriental hornet which is of approximately the same size, colour and shape, and which can be most savage when its nest is disturbed, or sometimes even when it is only approached.

The Little bee, *Apis florea*, also builds a single comb, but only a small one about the size of the palm of a man's hand, which it suspends from the branches of bushes or small trees. (Pl. II, p. 34). The combs of colonies of the Little honeybee contain about 100 worker cells per square inch. Their colonies only gather small quantities of honey, an ounce or two at most, but this honey is highly esteemed for medicinal purposes by many people in whose countries it is obtained.

These little bees are very gentle and seldom attack man even when their nests are disturbed. When they do attack they have great difficulty in forcing their tiny, delicate, stings into the skin. The sting is almost as severe as that of any of the other honeybees; in fact, in the author's experience, there is very little difference in the intensity of the pain produced by stings from any of the four kinds of honeybees, although, no doubt, the smaller species inject less venom than the larger ones.

In the wild state, both Eastern and Western bees (*A. indica* and *A. mellifera*) build nests consisting of several, vertically hanging, parallel combs of various sizes in hollow trees and in cavities in rocks and walls, sometimes constructing as many as eight or ten combs. The sizes of these are adjusted by the bees to suit the shapes of the nesting cavities.

These two species are the hive-bees of the world and have for many hundreds of years been successfully kept by man in old cooking pots and similar vessels, and in such primitive hives as hollowed-out logs, tubular mud containers and straw skeps. Today, although a great many colonies are still housed in such primitive shelters, many others are, of course, kept in hives so constructed that their owners can handle the bees very easily and obtain much larger crops of honey than is possible with less adaptable equipment.

This habit of *A. indica* and *A. mellifera* of building their nests in sheltered places undoubtedly helped their dispersal throughout widely separated parts of the world and their great success; a success furthered by their habit of living on reserve stores of food accumulated during favourable weather conditions, which gives them a remarkable ability to withstand long periods of adverse conditions (the heat of tropical summers and the cold of European winters) when fresh nectar and pollen cannot be obtained in the field. The Giant bee and the Little bee with their exposed combs have never been able to compete with these bees, except in the tropics, and have not been able to colonise the colder parts of the world.

Europe, Asia and Africa were, almost certainly, all joined together to form one great continent during the period when the colonies of the ancestors of our own Western honeybee were dispersing from their home in South Asia, so that there were no oceanic barriers to the spread of the hive-bee throughout these vast areas of the world. This great continent was, however, separated by sea from the American continent with the result that no honeybees reached America until taken there by man in recent times.

The Western honeybee is said to have been imported into America from Europe as early as the middle of the seventeenth century, but how and whence it came are most uncertain. It scarcely seems possible that colonies of honeybees could have been successfully taken from Europe to America in sailing ships which took so long to complete the voyage, unless, perhaps, the voyage was made during the winter months at a time when the bees were clustering tightly on their combs and would be less inclined to attempt to fly. We really do not know how bees were successfully taken in the first instance on the long journeys from Europe to America and also to Australia, which again has no indigenous honeybees; but there is no doubt that colonies of honeybees made these journeys safely more than once.

As the hive-bees, *A. mellifera* and *A. indica,* gradually colonized the greater part of the world a number of races, or varieties, began to develop. During recent times geographical boundaries, such as oceans, deserts and mountains, have tended to keep some of these races more or less apart so that each has tended to become distinct from the others. Some of these races are very important today and are arousing a great deal of interest amongst beekeepers, as selective bee breeding becomes more widely practised. Fortunately, from the point of view of the bee breeder who desires to combine the best characteristics of several races of bees, members of even the most dissimilar races of *A. mellifera* will readily and successfully interbreed with one another.

A most striking thing about *A. mellifera* is the small amount of variation shown by bees of the different races. Even in extreme cases the sizes of various parts of the body, such as the tongue, differ very little.

During the last few decades, the great increase that has taken place in the speed of transportation, coinciding as it has with the growth of man's recognition of the economic possibilities of the honeybee, has had a profound influence on the distribution of the races of *A. mellifera*. Bees of a number of distinct races have been imported into many countries in ever-increasing numbers and have subsequently interbred with one another as well as with the indigenous population. In order to introduce a new race, it is only necessary to import a single fecundated queen accompanied by a dozen workers in a small cage weighing a few ounces, and such importations have taken place very rapidly into Britain and other countries, resulting in a considerable degree of interbreeding between the races and the production of racially hybrid and often inferior bees.

Until the introduction of the technique of instrumental insemination (Pl. 37, p. 198 and Pl. 38, p. 199) the maintenance of a pure strain of bees has been very difficult in most areas, and it is in consequence probably true to say that very few strains of honeybees with well fixed characteristics (pure strains) are to be found in Europe today. Since the last war, however, the Rev. Brother Adam and several other people have discovered a few of these pure strains still surviving in remote and relatively inaccessible places in parts of Europe and the Middle East.

Today two subspecies of *Apis mellifera* are recognisable, each with a number of more or less distinct races. These are:—the Western honeybee (subspecies *A. m. mellifera*) and the African honeybee (subspecies *A. m. adansonii*).

Some workers have recognised further subspecies, but the

a. Queen and workers of various sizes on brood clump. The first workers of the colony, reared by the queen herself, are very small. These workers forage for and rear subsequent workers and, as food becomes more readily available, larger adults are produced. (x 1·5)

b. Nest showing 'honey-pots' in which nectar is stored. (x 2·0)

Plate 3.—Bumblebee Nests

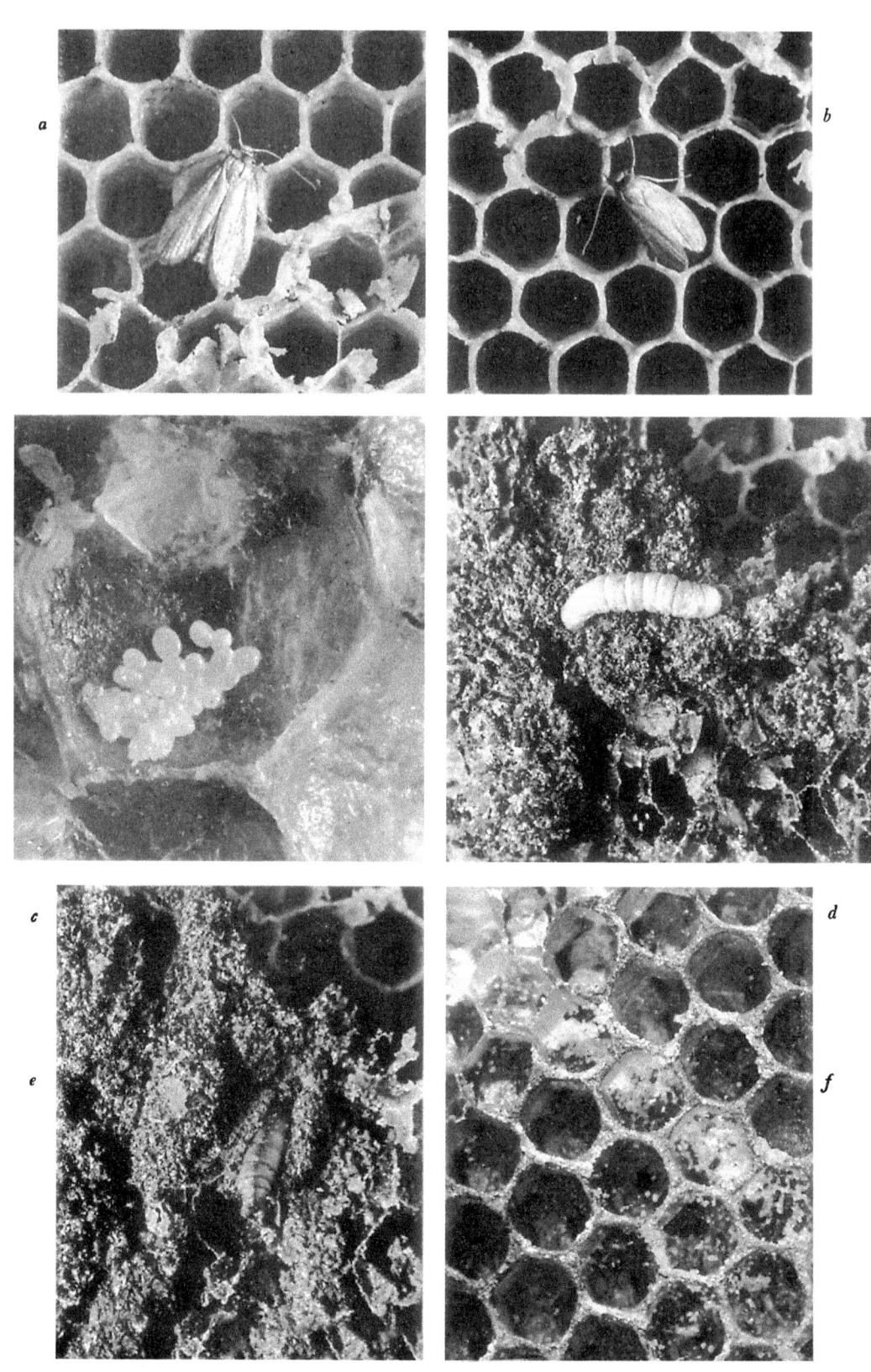

author prefers to regard those that are additional to the ones listed above as races rather than subspecies. I am aware that the word "race" is used by some systematists to mean subspecies; but here I use it to represent a lower category. It is, of course, only a matter of opinion where one draws the line.

The five most important races of the Western honeybees are:—

1. *lehzeni*, a dark, brownish-black bee found in Norway, Sweden, North Germany and in Britain—at all events until recently.

   It possesses a shortish tongue compared with some other races and is very hardy.

2. *mellifera*, the southern form of *lehzeni*, which possesses similar characteristics in a rather less well pronounced form. It occurs throughout central Europe north of the Alps. Colonies of this race are to be found in the Netherlands, Belgium and northern France.

3. *carnica*, the Carniolan bee, whose short greyish-white body hairs give it a most attractive appearance. This race possesses a decidedly longer tongue than the two preceding ones. Its chief habitats are in Austria and in parts of the Balkan countries. It is a very hardy bee, a diligent worker and a very successful defender of its home against robber bees and wasps.

4. *ligustica*, the well-known Italian honeybee, which is distinguishable from all other races by the beautiful yellow bands on the workers' abdomens. Its home is in the Italian peninsula from whence it has been widely distributed in large numbers to many parts of the world including Britain, America, Australia, and New Zealand. It is not as hardy under British conditions as some of the more northern races.

Plate 4 (opposite). LIFE HISTORY OF THE LESSER WAX MOTH
The wax moths invade the nests of bees in tropical and subtropical countries and do great damage to the combs upon which the larvae feed. Only strong colonies will keep them away. In Britain the nests of weak colonies are sometimes invaded but most damage is done to combs in store.
a. Adult female (x 2.6). *b.* Adult male (x 2.6). *c.* Eggs on comb (x 11). *d.* Full grown larva (x 2.2). *e.* Cocoons and pupa in mass in a damaged comb (x 2.8).
f. Silken tube of larva in comb. By making and living in such tubes the larvae are protected to some extent from attack by the bees (x 2.8).

5. *caucasia*, the Caucasian honeybee. This race has its home in parts of the Caucasian mountains and has an attractive grey colour and a long tongue. Many, if not all, of the strains of this race that have been tried in Britain have shown an unfortunate tendency to collect a large amount of resin (propolis) from various plants and to use it to glue the interior fittings of their hives together. In autumn, bees of these strains will in many cases build curtains of this propolis, with a few pop-holes in them, right across the entrances of their hives.

Before about 1917 the so-called British Black or Brown bee (*Apis m. mellifera* race *lehzeni*), the indigenous race, was common throughout the British Isles, although hybridisation had taken place to some extent in some localities into which a few colonies of bees of other races had been imported. However, very severe losses of adult bees, often leading to the death of their colonies, were reported in the neighbourhood of the Isle of Wight in 1905 and, during the next few years, throughout Britain. These losses were believed to be caused by a disease of adult bees which became widely known as "Isle of Wight disease". The cause was unknown but later attributed to a parasitic mite, *Acarapis woodi*—hence the alternative popular name "acarine disease"—to which the British Brown bee was supposed to have been particularly susceptible. According to the late Mr. William Herrod-Hempsall, "Isle of Wight disease" first appeared on the mainland in 1909 in those counties nearest to the Isle of Wight, namely Dorset, Hampshire, Surrey and Sussex. It then appeared, according to the same author, in Cornwall, Glamorgan, Herefordshire and in the Western borders of Worcestershire and Yorkshire, as well as in Scotland and, later, Ireland. It was suggested that perhaps the importation of infected bees into these areas from those counties in which severe losses had occurred in 1909 was responsible. Alternatively, it was thought that perhaps the local conditions favoured the development of "Isle of Wight disease" in these areas in which it had probably been present in a mild form for some time. At all events it seems clear that widespread loss of bees occurred throughout Britain.

By means of a survey what was then the Board of Agriculture

estimated that the loss of colonies of bees in Britain by 1918 had reached the appalling figure of ninety per cent of the total population. It should however be realised that this did not mean (as has sometimes been suggested) that our indigenous Brown bee had been practically exterminated. There were many colonies of bees in Britain when the serious losses began and, even if ninety per cent of these were destroyed—as the results of the 1918 survey suggest—there must have still been many hundreds of colonies of British Brown bees left in different parts of Britain. Colonies of honeybees multiply rapidly by swarming, and the British Brown bee was reputed to swarm frequently. Indeed, the method of skep beekeeping that was still widely practised at this time encouraged this habit, so that it is probable that the population of bees would have recovered within a few years without any help from outside had it been allowed to do so. However, the tragic loss of bees was considered to be so serious that a Government-controlled importation scheme for restocking Britain with bees was put into operation in 1919 immediately after the first World War. The colonies were distributed from selected centres established for the purpose and most of them were headed by queens of the Italian race of the Western honeybee (*Apis m. mellifera* var. *ligustica*). Whether or not such importation and restocking with a race of the Western honeybee which was foreign to Britain (particularly with a southern race), was a wise policy will, I suppose, remain a subject of debate among beekeepers for many years to come.

Thanks to the careful researches of my colleague, Dr. L. Bailey, it now appears most improbable that the heavy loss of colonies in Britain—during the so-called "Isle of Wight disease" epidemic—was caused by any one infectious disease, such as acarine, or even by a group of diseases, although adult bee diseases no doubt played their parts. Indeed, as Dr. Bailey has pointed out, it was assumed that the cause of the losses was an infectious disease without any evidence and this assumption was then promulgated in sensational but uninformative articles which helped to fix the belief firmly in the public mind. It is very probable that starvation was a major cause of the deaths of many of the colonies.

Since the Government-sponsored importation scheme for restocking Britain with honeybees in 1919-20 a large number of colonies and queen bees of various European races, and even of some North African races, have been imported into Britain. These are mainly of French, Dutch and Italian origin, but include some Caucasian, Cyprian and Carnioloan bees, amongst others. Importation necessarily ceased soon after the beginning of the second World War, in September, 1939, but began again in 1947, although on a much smaller scale and in a more or less controlled way.

Since honeybees mate on the wing or at any rate in the open air, and as isolation of queen-mating apiaries is quite impossible in most of the beekeeping areas in Britain, a great deal of racial hybridisation has taken place in an entirely haphazard manner. During this hybridisation the good qualities possessed by some of the imported bees have tended to become lost, and it is a sad fact that the great majority of honeybees in Britain today are of hybrid and inferior stock.

CHAPTER 3

THE EVOLUTION OF
SOCIAL LIFE AMONGST BEES

A SOCIAL INSECT has been defined as "an insect which lives in a society; each society consisting of the two parents, or at least the fecundated female, and their offspring, and with the two generations living together in a common abode or shelter and exhibiting a variable degree of mutual co-operation". All true insect societies are in fact families. This being so, we must conclude that insect social life in any highly developed state, such as that found in the honeybee colony, has evolved slowly through numerous stages. These stages show an increasing degree of intimacy between the mother insect and her offspring, and, with it, an ever-increasing degree of mutual co-operation. Any idea that the societies of ants, bees and wasps may have arisen from chance associations of female insects of the same species seems most unlikely.

Apart from the temporary association of a male with a female for the purpose of mating, the vast majority of insects live independent solitary lives. There are, however, a number of species living today which exhibit a tendency towards social behaviour, as earlier defined. It is amongst these *subsocial* species that we can find a graded series of behaviour patterns demonstrating the probable evolutionary steps which have culminated in the highly developed social organisation of honeybee, ant and wasp colonies. Here it must be emphasized once again that mere gregariousness probably had nothing to do with the evolution of social life in insects, as will be realised from the rather unsatisfactory definition of social life which has been quoted. Examples of gregarious insects exhibiting no social life are certain caterpillars (e.g. the communal webspinning caterpillars of the Lackey moth) and, of course, locusts. Any truly social insect association is based fundamentally upon the family, as are so many animal societies, including that of man.

True examples of social life amongst insects are found only in a very few species of bees and wasps, in the ants, and in the termites or so-called "white ants"; that is, in only two Orders, the Hymenoptera and Isoptera: though examples of subsocial life are to be found in a number of other Orders of insects.

The Hymenoptera and the Isoptera are not at all closely related to one another. The Isoptera consist of primitive insects related to the cockroaches, whereas the Hymenoptera embrace very highly specialised insects. None the less, the organisation of life in the termite colony is in many respects closely parallel to that found in colonies of honeybees, although it must have originated entirely independently.

Let us now attempt to trace out the probable steps followed in the course of the evolution of social life, from the purely solitary life shown by the great majority of insects to the highly organised state of affairs found in the honeybee colony.

We can start with a solitary insect such as the little primitive bee *Prosopis pectoralis*. Males and females of this bee are present in marshy areas in Britain in early summer in approximately equal numbers, and the sexes co-operate with one another only to the extent of mating together. Soon after she has been fecundated each female bee seeks out, quite independently, a nesting site in an old, disused, gall of the Reed Gall-fly, *Lipara lucens*. This gall consists of a cigar-shaped aggregation of dead leaves at the apex of a flowering-stem of the common reed, *Phragmites communis*. In the lumen of the gall the female *Prosopis* commences to build her cells one at a time. Each cell is first lined with a thin, translucent, membrane and the little female bee places in it a rather liquid paste consisting of pollen and nectar which she regurgitates from her crop. She possesses no pollen-baskets nor other special pollen-collecting apparatus so she swallows the pollen along with the nectar when collecting them from flowers. Having provisioned a cell in this way she lays a single egg on top of the pellet of pollen-nectar paste, and immediately closes up the cell with the thin, translucent, material already mentioned. She then proceeds to construct and provision another cell above the first one, lay an egg in it and seal it up. She continues in this way until the lumen of the gall is occupied by her nest of five to eight cells. Then, worn out, she dies.

The larvae, which hatch from the eggs within a few weeks, feed upon the food so carefully provided in their own cells and have no contact with one another. Eventually, by the time the autumn arrives, the larvae are all full grown and they remain thus in a dormant state until the following spring when each pupates in his or her own cell. Each perfect insect emerges in May or June, quite independently of the others, and each little bee then leaves the parental nest and seeks a member of the opposite sex with whom to mate. After mating the females build nests of their own and thus the whole cycle is begun all over again.

The male *Prosopis* dies soon after mating, his life's work complete, and the female with whom he has mated never has any contact with her offspring, in fact she does not live long enough to see them. She practises what is known as mass-provisioning by providing a lump of pollen paste about the size of a small pea in each cell. This is sufficient food for the needs of the larva which will hatch from the egg which she lays on the paste.

Contrary to popular belief most of the bees, not only of Britain but of the world, are solitary in habit, and have life-histories very similar to that of *Prosopis pectoralis*.

The first stage in the development of social life from a solitary life is exhibited by the Pentatomid or shield-bug, *Elasmostethus griseus*, which lives on birch trees in Britain. The female shield-bug lays a batch of eggs on the underside of a birch leaf and proceeds to hide them under her body until the nymphs have hatched from them. She then continues to guard them for a few days. Here, in the case of *Elasmostethus griseus* we see that the length of life of the female has been prolonged sufficiently for her to see her offspring and, indeed, to show some care for them; but it should be noted that there is no sign of mutual co-operation between the mother and her offspring and no common abode or shelter is used. This then is a very rudimentary case of subsocial behaviour.

The earwig, *Forficula auricularia,* which is so commonly found seeking shelter inside the roofs of bee-hives, exhibits a slightly more advanced stage of subsocial behaviour. In this case the female guards her eggs, and the nymphs which emerge from them, in a shelter which she finds ready made or else constructs in the ground, or in the hollow stem of some plant.

A similar state of affairs to this is found among certain species of small bees belonging to the genus *Halictus,* which belongs to the family Andrenidae. The nest of *Halictus quadricinctus,* for example, is built in a burrow in the ground, and its cells are *progressively* provisioned by the mother during larval development and ultimately sealed up. The mother, however, continues to survive and to guard the nest until her offspring emerge. She then dies and her offspring disperse. In this case progressive feeding of the larvae takes place, similar to that which occurs in the honeybee colony. But there is no mutual co-operation between the mother and her offspring.

A much more advanced stage in development towards full social life is found in another small British bee, *Halictus malachurus.* Large numbers of these little bees consort together in favourable localities and *individually* excavate nesting burrows close together in hard soil. One frequently hears the name "colonies" applied by naturalists to such groups of solitary bees' nests, of which there may be hundreds very close together but quite distinct from each other. Such "colonies" however, are merely aggregations of nests close together in favoured areas and must not be confused with the colonies of social bees.

Early in the spring, when the willows and dandelions from which they collect much of their food are in flower, the over-wintered females of *Halictus malachurus* begin to appear. These females were fertilised during the previous autumn. Although several females usually overwinter together in a single burrow quite peacefully, they now fight amongst themselves until only one is left in possession of the burrow. Each of the ousted individuals must get a new burrow for herself. After each female has acquired or dug one, she proceeds to build her brood cells in short passages which she excavates at the sides of the main burrow or gallery. As happens with most of the other species of *Halictus* and *Andrena,* as soon as each cell is built it is mass-provisioned with a pellet consisting of a mixture of pollen and nectar that the bee has collected. Unlike that of the primitive little bees of the Family Prosopidae, however, this pollen-nectar pellet is relatively dry and of the consistency of dough. An egg is laid on each pellet as it is completed and the cell in which it is contained is sealed over. The mother pays no further attention to this cell, but continues to build, provision, and lay eggs in further cells.

By the end of May those larvae which hatched from the earliest eggs to be laid are full-grown and pupate, and by mid-June the first of them begin to emerge from their cells as perfect insects. These adults are all females. They are somewhat smaller than their mother and have different markings on the cuticle of their bodies. Before their origin was discovered, they were considered by taxonomists to belong to a different species which for many years was called *Halictus longulus.*

The old mother *malachurus* still remains alive and her *longulus* daughters remain with her in the nest. They do not mate, however; indeed no males are available with whom they could do so. Instead they immediately make further brood cells and collect nectar and pollen in the field with which to provision them; but they do not lay eggs in them. This continues to be done by the old mother *malachurus* alone. She does not go out foraging any more once the first few of her daughters have appeared, but instead devotes her time to guarding the entrance to her nest-burrow, and to laying eggs in the cells as they are prepared and provisioned by her *longulus* daughters.

By the beginning of August each nest contains an old mother *malachurus* and about a dozen of her *longulus* daughters. About this time *malachurus*-type females and also some males begin to appear, being produced, like the *longulus* females, from eggs laid by the old mother *malachurus.* During fine sunny weather the males mate just outside the nest-burrows with young *malachurus*-type females both from their own and other burrows; but pay no attention to the *longulus* females. The latter continue to collect pollen and nectar and to build and provision further cells. The new *malachurus* females visit flowers in search of food for themselves but they do not bring any food back to the parental nest, in which they continue to live; they do not help in the economy of their colony in any way, as far as we know.

When colder weather arrives in the autumn the old *malachurus* mother dies, as do the *longulus* females and the males. The young, impregnated, *malachurus* females live on, however, and proceed to hibernate in the nest in which they were born until they emerge during the following spring, to found nests of their own and to play their part in repeating the whole cycle of events.

Clearly *Halictus malachurus* exhibits a well-marked step towards the development of a truly social organisation. Not only does each nest contain a single laying *malachurus* queen but also a small number of workers—sterile females—for this is what the *longulus* females are. As soon as these workers have been produced they take over from their mother the duties of foraging for food and building the nest. She concentrates her energies on guarding the entrance to the nest burrow and producing further eggs— "stealing" the food she requires from the pollen-nectar pellets which her daughters are building up in newly prepared cells. On the other hand the cells are all mass-provisioned and there is no progressive-provisioning, as practised by the most highly developed social insects, the ants, social wasps, honeybees and bumblebees, etc. The nest remains a relatively simple structure similar to that produced by many of the solitary bees and wasps.

Several other species of the genus *Halictus*, such as *H. maculatus* and *H. immarginatus*, also possess a subsocial colony organisation similar to that exhibited by *H. malachurus*.

These three species of bees, *H. malachurus*, *H. maculatus*, and *H. immarginatus*, have, as pointed out by Dr. Wheeler in 1928, developed a form of social life that is almost as advanced as that of the truly social wasps and bumblebees. The sterile females found in the nests of these three kinds of bees during the summer months correspond, of course, to worker wasps and worker bumblebees, and the mother bee represents the overwintered wasp or bumblebee queen. There is, however, no evidence of progressive-provisioning of the cells, that is feeding of the growing larvae only as and when they require it. Although this does not occur it has been found that some of the empty brood cells of a related species, *H. scabiosae*, are sometimes used for the temporary storage of pollen.

A somewhat different step towards the evolution of social life is illustrated by some (otherwise) primitive little South African bees of the genus *Allodape*. These species of *Allodape*, which were studied by Dr. Brauns in 1926, build no cells at all and their nests consist merely of the hollow stems of plants. Each nest is built by an individual female who lays her first egg at the top of her nesting cavity, attached to the side, and not on a pollen-nectar pellet at all. She then proceeds to lay further eggs and to attach them at intervals along the walls of her nesting cavity in the hollow stem. As soon as an egg hatches she feeds the resulting larva with nectar and pollen. The mother *Allodape* feeds the larvae one at a time, in some species by mass-provisioning, supplying at one time all the food that a larva will require to reach maturity, but in others by progressive-provisioning. In the latter cases each larva is given only a small quantity of food as and when it requires it. All the females produced in *Allodape* nests are like their mothers and are believed to share in the feeding of the larvae. However, these young females soon leave the nests, in which they were hatched, to mate, and after mating they quickly set about founding new nests of their own. They continue for some time, however, to return to their original homes at intervals and to assist with the feeding of their brothers and sisters. Here then we find cases of progressive feeding of larvae and the co-operation for a short time of a number of "normal" females with their mother in feeding the larvae in the parental nest. I have called these females "normal" as they are identical with the mother, capable of mating and not in any way differentiated as workers.

Another step towards a truly social life as exemplified by the honeybees, has been taken by the bumblebees. The colonies of all our British species are of an annual nature only and each is founded in late spring or early summer by a single fecundated queen. She was mated in the previous summer or autumn and has overwintered without feeding, in a hole in a bank or in some similar situation amongst moss and dead leaves. Each individual queen builds a small nest consisting of only a few cells. The nest of some species is built in a hole in the ground and that of others in the middle of a tuft of matted grass, often in the discarded nest of a mouse, or some such place. The cells are built of wax secreted by special glands of the queen's body. Except for one or two special "honey-pot" or storage cells (Pl. 3b, p. 18), which are the first to be built, each cell receives several eggs. The larvae which hatch from these eggs are fed progressively by the queen on nectar and pollen which she herself collects in the field.

The bees which emerge from these first cells are all sexually undeveloped females or workers, which are incapable of mating— indeed there are no males available with whom they could mate. These workers are considerably smaller than their mother, the queen (Pl. 3a, p. 18). Some of the first of them to emerge are very tiny indeed, having only enjoyed a near-starvation diet. A few hours after they have emerged from their cells the workers take over from their mother the duties of nest-building, foraging for food and the nursing (feeding) of her larvae. The queen now becomes little more than an egg-laying machine.

Towards the end of the summer a number of large female bumblebees which are sexually mature are produced, and so are a number of males. These virgin queens and the males, or drones, leave the nests in which they were reared, in search of food and mates. As the young queens and the males of any given colony are usually produced at slightly different times, mating of brothers and sisters is not the general rule; the young queens of one colony usually mate with drones from another.

The queens and drones continue for a time to use the nests in which they were reared, as shelters, but play little part in the economy of their colonies.

Ultimately each colony breaks up; the old queen, males and workers die, and the young impregnated queens go off individually to seek places such as holes in banks in which to rest quietly until the following year.* The following spring these overwintered queens leave their resting places on a warm day, each to attempt to found a colony of her own.

*Such quiescence of impregnated bumblebee queens has often been called "hibernation" which implies that the winter has something to do with it. In this connection it is, perhaps, as well to remember that the colonies of some species of bumblebees (such as *Bombus pratorum*, the queens of which appear early in the spring and which are the first of our British bumblebees to found

EVOLUTION OF SOCIAL LIFE

Among bumblebees, then, there is a marked differentiation between the queen and worker castes, similar to that found in *Halictus malachurus*. Progressive feeding of the larvae is usually carried out, in the same kind of way as occurs in some species of *Allodape*, with a small amount of food storage, as in the honeybees. But the queens can, and do, perform all the functions that are later carried out by their workers once these have appeared. Bumblebee colonies are only small annual affairs relative to honeybee colonies: a large colony of *Bombus terrestris*, a prolific species, only contains about 400 workers as compared with perhaps 50,000 workers of a large colony of *Apis mellifera*. Bumblebees do not swarm, and only store enough food to tide the colony over a very few days of bad weather when foraging is impossible. There are, however, reported to be a few sub-tropical species of bumblebees which form perennial colonies, each of which contains a number of fertile queens and from which dispersal takes place by means of small swarms.

In the *Meliponinae* or so-called "stingless" bees, many species of which are found in South America, caste differentiation is well marked and the queens have lost all trace of pollen-collecting apparatus. The workers are sexually undeveloped, do not mate, and are often present in large numbers. Those of many species are extremely small (Pl. 2a, p. 3). These bees build their nests in hollow trees, holes in walls, and sometimes even in empty beehives. The larvae are reared in individual cells or cups which in some species are grouped together in a different part of the nest from that in which the food storage cells (for pollen and nectar) are found (Pl. 1, p. 2 and Pl. 2, p. 3). The colonies are perennial and swarming takes place from time to time. The social organisation of colonies of Meliponine bees is very like that of honeybee colonies, except that the larvae are fed by mass-provisioning rather than by progressive-provisioning of their cells. These little bees occupy a position that is more or less intermediate

colonies each year) actually begin to break up during the height of summer. Thus impregnated queens of *B. pratorum* have been found "hibernating" during some of the warmest times of year. It is perhaps wiser, therefore, to refrain from using the term "hibernation" when referring to the quiescent stage in which many queen bumblebees remain from summer until the following spring.

between that of the bumblebees and the honeybees; and show the most highly developed social life we have so far studied.

We come now to the true honeybees; the Giant honeybee, *Apis dorsata;* the Little honeybee, *Apis florea;* the Eastern honeybee, *Apis indica;* and the Western honeybee, *Apis mellifera.* *A. dorsata* has queens, drones and workers, the workers and drones being reared in cells of similar shape and size. Colonies of *Apis florea* also nest in the open air, like the Giant bees, but differ from them in building different kinds of cells in which to rear their queens, workers and drones—thus resembling *A. indica* and *A. mellifera.*

It is in colonies of *A. mellifera* and *A. indica* that we find the highest degree of social development amongst the bees. The composition of these colonies and the caste differentiation involved will be discussed later on (Chapter 4, p. 34), but we may say here that a very distinct worker caste is present consisting of sterile females and that each colony normally only possesses a single queen who is the mother of all the bees forming her colony. Queens of all four of the species of honeybees (*Apis sp.*) have no trace of pollen-collecting apparatus; wax-secreting and special scent-producing glands and their hypopharyngeal glands are also vestigial. These queens have become immensely fertile and capable of producing many thousands of eggs in the course of their lives, which may last for several years.

The colonies of all true honeybees are perennial and the development of new colonies takes place by swarming. At the height of its strength in the summer a good colony of the Western honeybee may consist of a single queen, a few hundred drone or male bees, and as many as 50,000 or more worker bees, together with their combs containing brood and food. Colonies of the Eastern honeybee tend to be composed of slightly smaller numbers of individuals, as do those of the Giant honeybee; and colonies of the Little honeybee are usually very much smaller and only contain some 4,000-5,000 individuals.

Thus we see that amongst bees living today, it is possible to trace a series of social levels, or steps extending from the simple, solitary, type of life, through a number of subsocial forms to elementary but truly social forms, and on to the most highly organised social bees of all, the honeybees *Apis mellifera* and *Apis indica*. A series of very similar steps can also be traced amongst the wasps. Nobody can, of course, say that these steps are exactly the same as were the evolutionary steps which led to the development of the highly organised social life of the Western honeybee; but it seems highly probable that the ancestors of our honeybees did follow some such evolutionary path.

CHAPTER 4

THE ORIGIN OF
THE MEMBERS OF THE COLONY

"It hath been an ancient tradition, that bees have a twofold production, as many other creatures, by generation and putrefaction. But whether there be any solidity in this particular, I am somewhat dubious. . . ."

> SAMUEL PURCHAS (1657). *A Theatre of Politicall Flying-Insects*

THERE IS no doubt that the honeybee has been kept for profit by man for several thousand years at least. It is simple to imagine how beekeeping gradually took the place of bee-hunting and evolved throughout the centuries into the art as we practise it today. One can imagine how some of the honey-seekers grew tired of fighting their way through the forests in order to locate and rob the nests of wild bees, in the way that some primitive peoples still do to this day, and experimented by cutting out those sections of hollow trees in which the bees were living and setting them up in conveniently accessible places. But, alas, we do not know during what era of man's history this first occurred. Indeed, as Dr. H. M. Fraser has put it, "the beginning of the art of bee-keeping is lost in the mists of antiquity".

It is clear, however, from the writings of Aristotle, Virgil, and others, that for about two thousand years some beekeepers have been aware that during the summer months every colony of honeybees is composed of a large number of individuals of which three different kinds can be distinguished. These are, of course, what we know today as the queens, drones and workers. (Pl. 6, p. 39).

Plate II.—The Little Honeybee (*Apis florea*)
This honeybee lives in South Asia. Like the giant honeybee its colonies build
their single-comb nests in the open air. A large nest is about the size of the
palm of a man's hand. The one shown was only about half this size and was
attached to a twig of a large bush. (x 1·7) (*see also Plate 10, p. 67*)

Early writers on the natural history of honeybees devoted considerable space to speculation on the origin of bees in general and of these three kinds of bees in particular. Aristotle, who was one of the very few early writers to attempt to distinguish between what had definitely been proven and what was only conjecture, made no claim to understand the generation of bees and only tells us that "some people say that bees collect their young from certain kinds of flowers." Virgil seems to have been prepared to believe that bees are generated in this rather surprising way, but he also repeats the fable, which was widely believed both in his own day and subsequently, that "bees can arise from the corpses of oxen that have been carefully killed without any blood being shed". Servius was one of the first to attempt to explain separate origins for drone and worker bees and Dr. Fraser quotes him as writing that "bees are said to be derived from oxen, drones from horses, hornets from mules, and wasps from asses."

Many of the early writers believed that those bees which we now describe as queens were kings. They did not definitely know the sex of any of the three kinds of bees that are to be found in a honeybee colony. It was not until the Reverend Charles Butler of Magdalen College, Oxford, published his justly famous book *The Feminine Monarchie* in 1609 that the sex of the drones, the male bees, was established and, simultaneously, the statement made that the queens are females. Butler made the mistake, however, of believing that the workers mate with the drones and lay the eggs from which both workers and drones are produced.

Charles Butler's description of the drone is well worth quoting. "The drone, which is a gross hive bee without sting, hath been always reputed a greedy lazer. For howsoever he brave it with his round velvet cap, his side gown, his full paunch, and his loud voice; yet is he but an idle companion, living by the sweat of others' brows. For he worketh not at all, either at home or abroad, and yet spendeth as much as two labourers; you shall never find his maw without a drop of the purest nectar. In the heat of the day he flieth abroad, aloft and about, and that with no small noise, as though he would do some great act; but it is only for his pleasure, and to get him a stomach, and then returns he presently to his cheer." I am afraid that the Rev. Charles Butler clearly

D

disapproved of drones and was scarcely fair to them, and I sometimes wonder whether he was not perhaps largely responsible
for the slanderous epithet "as idle as a drone".

In 1637 Richard Remnant in his *Discourse or Historie of Bees*
rightly concluded (but on incorrect grounds) that worker honeybees are females because each possesses "a neat place for the
receipt of generation."

Various other writers on this subject came to the conclusion
that the drones are males and fertilise the eggs (which some, such
as John Thorley (1744) correctly believed to be laid by the queen
bee) in much the same way as many male fishes eject their milt
over eggs newly laid by the females. In 1771, however, a full
and extraordinarily accurate description of the process of swarming, and of the mating of the virgin queen outside her hive and
how she thus becomes the mother of a colony was actually
published by Anton Janscha, the son of poor country folk, who
had knowledge of both theoretical and practical beekeeping in
advance of his times and became the Imperial and Royal Beekeeper to the Empress Maria Theresa herself.

Anton Janscha's two books, in which he mentions his discoveries about the life history of the honeybee, seem soon to have
been forgotten. Many educated people continued to believe in
the theory of spontaneous generation of life until well into the
nineteenth century, so it is scarcely surprising that mythical stories
of the generation of bees from oxen, and so forth, continued to be
believed from classical times until the nineteenth century. By
the beginning of this century, however, the work of the great
Swiss naturalist, François Huber, on the life-history of the honeybee—some of which had, unknown to Huber, been discovered by
Janscha twenty years earlier—was becoming more widely known
amongst beekeepers, and the old mythical stories to explain the
generation of bees soon disappeared.

Today it is universally recognised that queen and worker
honeybees are both females, and that the drones are the males.
It is also well known that during the summer months an average
sized colony of honeybees is composed of thirty or forty thousand
worker bees, a few hundred drones, and a single queen bee who
is normally the mother of all the other members of the colony

to which she belongs. Now, as their names imply, the two kinds of females found in every normal colony of bees—the queen and her workers—exhibit a very marked division of labour between them. The queen honeybee alone is responsible for the production of the eggs from which all the members of the colony arise—workers and drones and, under exceptional circumstances, further queens. As far as is known this is the only job done by the queen honeybee for her colony. On the other hand, the other kind of females, the workers, do not as a rule lay any eggs at all, although they can become capable of so doing in exceptional circumstances (p. 57). Instead, they are concerned with all the other everyday requirements of the colony, both in the hive and out, including the whole production of future generations except the actual egg-laying. In every normal, queen-right,* honeybee colony we find a perfect co-operative partnership between all the females present, the queen producing all the eggs and leaving the care of them, and of the larvae which hatch from them, entirely to the worker bees. This does not mean, however, that the queen has only a light task, for in a prosperous colony during the early summer she will produce about 1500 eggs every day. This is more than her own body-weight in eggs every twenty-four hours. For the purpose of egg-laying, and for this purpose only, the queen honeybee has become highly specialised. She has lost certain structures that are possessed by the workers and are necessary for carrying out their many and varied duties, but which are unnecessary for egg-laying, for example the long tongue for collecting nectar from flowers, the pollen-baskets in which to carry pollen home, and wax- and brood-food-producing glands.

The queen honeybee has become so highly specialised as an egg-producing machine that she is quite unable to collect food for herself or her offspring, or to build a nest (or any part of it). She cannot even survive for very long on her own. In this way the queen honeybee differs from the queens of British bumblebees and wasps which, having overwintered individually in sheltered places, each set about the task of founding their own colonies. This task involves, amongst other things, finding a suitable nesting site, building a rudimentary nest, producing the eggs and

*A beekeeping term indicating the possession of a queen, usually a mated one.

incubating them, and providing enough food for herself and the first group of her offspring. Indeed the queen bumblebee or wasp goes on doing all these things until this first group of five or ten daughter workers have become adult, and are old enough to relieve their mother of all duties except those of egg-production and incubation.

It could be said that the queen bumblebee or wasp, being able to do all these things, is more highly developed than the queen honeybee; but this confuses versatility with specialised progress. The queens of the ancestors of our present honeybees were indeed able to do all these things; but their species' present success lies in the specialisation of their queens in egg-production to the ex-clusion of other duties. For whereas queen honeybees can produce as many as 1,500 eggs in a single day, few bumblebee queens, even of the more prolific species, produce as many in the whole of their lives.

We must now consider the question of how the three kinds of individuals, queens, workers and drones, found in honeybee colonies arise.

In 1845 a Polish beekeeper, the Rev. Johann Dzierzon, brought forward his now famous theory to explain the production of queen and drone honeybees. The gist of the Dzierzon theory is that when a queen honeybee has mated with a drone somewhere outside her hive and has returned to her colony again, she becomes able to lay two kinds of eggs. These two kinds of eggs differ from each other only in one respect: one kind of egg is unfertilised, the other kind is fertilised. The unfertilised eggs give rise to males, drones, only; whereas the fertilised eggs always give rise to females which can become either queens or workers.

Dzierzon's theory that parthenogenesis, or virgin birth, occurs regularly in the honeybee colony was hotly disputed for many years but is today fully accepted, having received abundant support from the data obtained in numerous experiments. For

Plate 5 (opposite). QUEEN WASP SEEKING NECTAR FROM THE EXTRA-FLORAL
NECTARIES ON THE SCALES OF THE BUD OF A CORNFLOWER
Wasps frequently visit flowers of the more open type in search of nectar, which appears to form an essential part of their diet (x 3.3)

a. Queen

b. Drone

c. Worker

Plate 6.—ADULT MEMBERS OF THE HONEYBEE COLONY
All x 4·5. Specimens chilled to reduce movement

example, if we mate together a queen of a pure strain of yellow-coloured American-Italian bees and a drone of a pure strain of Swiss Nigra bees which, as their name implies, are black, we find that all the worker-bees that are produced by this queen are yellow, like their mother, and that all the drones are also yellow. Now, if Dzierzon's theory is correct we can explain this result by saying that the drones are yellow because they inherit yellow body-colour from their mother and nothing (therefore no black colour) from any father, as they have no father. The yellow colour of the workers can also be explained on the assumption that, although each inherits a gene for black body-colour from her father, and a gene for yellow body-colour from her mother, yellow is dominant to black and, therefore, the black is suppressed and the resultant workers are yellow. But if (in order to explain the colour of the worker-bees obtained from this cross) we assume that this particular type of yellow body-colour is dominant over the Nigra's particular type of black body-colour, we must also be prepared to explain the yellow body-colour of the drones in the same way—supposing Dzierzon's theory is incorrect and that drones are also produced from fertilised eggs, i.e. have fathers and obtain characteristics from them as well as from their mothers. In other words, the results of this experimental mating, although of considerable interest in themselves, do not enable us to say whethers Dzierzon's theory is correct or incorrect; that is whether the drones arise from fertilised eggs or not.

However, when we mate one of the black Nigra queens with one of the yellow Italian drones, we find that all the resulting drones are black and that all the workers are yellow. This is precisely the result we would expect to get if Dzierzon's theory is correct. The workers are yellow, although they came from fertilised eggs and received genes both for yellowness and blackness, because this particular yellow is dominant over this particular black; and all the drones are black, despite the fact that yellow is dominant over black, because, coming from un-fertilised eggs, they received their characteristics from their mother only and therefore only the character of blackness.

Such evidence can be multiplied many times over; and this famous theory of Dzierzon has become the foundation stone of modern bee-breeding, although as we shall see, a few modifications have recently had to be introduced to allow for certain rare exceptions to the general rule.

Although we have known for a good many years now that female honeybees are produced from fertilised eggs and male honeybees from unfertilised eggs, the nature of the mechanism whereby this is brought about has been a matter of considerable dispute.

As long ago as 1913 a German worker, Nachtsheim, claimed to have succeeded in demonstrating that the unfertilised egg which develops into a drone possesses only sixteen chromosomes, whereas the fertilised egg which develops into a female honeybee (worker or queen) contains twice this number of chromosomes. All the chromosomes in the unfertilised egg are derived from the mother, whereas half the chromosomes in the fertilised egg are derived from the mother and half from the father—i.e. from a spermatozoon from the drone with whom the queen bee mated. The hereditable characteristics, such as body-colour, are, of course, carried on the chromosomes. In other words Nachtsheim and others have attempted to explain sex determination in the honeybee on a simple quantitative basis, the female sex having exactly twice as many chromosomes as the male sex. However, one or two other observers have suggested alternative mechanisms for sex determination, such as the possession by one sex only, usually the female sex, of a definite sex chromosome. They suggested that any egg which possessed this particular, unpaired, chromosome developed into a female honeybee, whereas an egg which did not possess this particular sex chromosome could only develop into a male. However, the whole question has been thoroughly reinvestigated by Sanderson and Hall who have been able to confirm Nachtsheim's view that sex in the honeybee is determined on a purely quantitative basis and that no special sex chromosome is involved.

In every normal colony of *Apis mellifera*, during the summer months, eggs and larvae are found in at least two kinds of cells, the so-called worker brood-cells and the drone brood-cells. These are of the same general shape but of the two the drone cells are somewhat larger—whereas there are on the average just under five worker cells per linear inch of completed comb, there are on the average only four drone cells per inch. The wax cappings with which the adult worker bees of a colony close the worker and drone cells when the contained larvae are full grown and almost ready to turn into pupae also differ somewhat in shape, those of the drone cells being more convex, more highly domed, than those which cover the worker cells.

At certain times, as will be discussed later, a third kind of cell is built by the worker bees. This is the queen cell, which it totally different in appearance from that of either worker or drone cells (Pl. 12b, p. 83) and is, as its name suggests, the special kind of cell in which a future queen may be reared. Queen cells are shaped somewhat like an acorn and are built so that their mouths hang downwards from the faces or edges of brood-combs, unlike the worker and drone cells, which are always arranged with the long axes almost horizontal.

We may thus recognise three kinds of cells in a colony of *A. mellifera*. In each cell only one larva is reared at a time, and whereas the same worker and drone cells are used over and over again to rear successive generations of bees, each queen cell is used once only and then destroyed by the worker bees.

The Giant honeybee (*Apis dorsata*) (Pl. 9, p. 66), the most primitive of the honeybees, rears its workers and drones in cells of similar size and shape except that just before the drone cells are capped the workers lengthen their side walls slightly so making the drone cells a little deeper than worker ones. Their queen cells are similar to those of other honeybees. The Little honeybee (*Apis florea*) (Pl. II, p. 18), however, rears its queens, workers and drones in three different kinds of cells which, apart from being much smaller, are in every other respect similar to those produced by the Western honeybee (*A. mellifera*) (Pl. 11b, p. 82) and the Eastern honeybee (*A. indica*) (Pl. 11a, p. 82).

Now, as we have already seen, it is necessary for the queen of a colony to lay unfertilised, drone-producing eggs, in drone cells, and fertilised, worker- and queen-producing, eggs in worker and queen cells, if the appropriate individuals are to be produced in the respective kinds of cells. This the fertilised queen honeybee does, very seldom placing the wrong kind of egg in a cell; although a queen which has been prevented from mating and so from receiving a supply of spermatozoa, or a queen whose supply has been exhausted, will lay unfertilised, drone-producing, eggs in worker cells.

A number of years ago the author watched a queen honeybee, whose colony was living in a glass-walled observation hive, laying eggs in a comb which was composed of numbers of worker cells with small groups of drone cells interspersed amongst them. She laid eggs in a number of worker cells in quick succession and immediately afterwards, without any more delay than if she were laying in further worker cells, she proceeded to lay eggs in three or four drone cells, then moved to more worker cells followed once more by further drone cells, and so on. In due course larvae hatched from these eggs and, when they were ready to pupate, their cells were sealed. This comb was then removed from the hive and all the adhering adult bees were shaken from it before it was placed in an incubator. When the time came for the young adult bees to emerge from their cells, it was observed from what kind of cell each of about fifty workers and an equal number of drones emerged. In every case it was found that a drone emerged from a drone cell and a worker from a worker cell, thus demonstrating that the queen had laid the appropriate kind of egg in each cell. How did the queen secure or inhibit the fertilisation of the eggs which she laid ? Unfortunately nobody seems to have been able to answer this question in any very convincing manner.

Let us consider the problems involved. When a queen honeybee mates she receives a supply of spermatozoa which the drone injects right into the inner parts of her paired oviducts, and the semen in which the sperm are contained is retained within the queen's genitalia by a mucous plug which hardens on contact with the air. This hardened mucous plug, to which remnants of the drone's genitalia are often attached, having been torn from him at the conclusion of the mating act, is usually displaced when she mates with another drone, or removed by the worker bees when she gets home. In the meantime, however, the sperm contained in the semen have moved towards the spermathecal duct and made their way up this tiny tube which leads from the median oviduct into a small balloon-shaped hollow body known as the spermatheca or sperm-reservoir. Here, possibly owing to being tightly packed, the sperm become quiescent and remain thus until some are released to fertilise an egg that is passing the mouth of the spermathecal duct on its way to the exterior. It is believed by many people that as each egg that is destined to give rise to a female bee passes the mouth of the spermathecal duct a muscular sphincter surrounding the mouth of the duct relaxes momentarily and allows a few sperm to pass into the oviduct. Once in the oviduct they become extremely active and apparently are strongly attracted by the egg and, on reaching it, one or more succeed in entering by means of a minute hole, the micopyle, at one end. As soon as one, or rarely more, sperm have succeeded in passing into the egg a special mechanism comes swiftly into play and effectively prevents the entry of further sperm. Thus the egg is fertilised and ultimately gives rise to a female honey-bee.

Now, let us consider what happens when an egg that is destined to become a drone is in the process of being laid. It must not be fertilised, therefore it must somehow pass down the oviduct to the exterior without a sperm entering it. How is this brought about? Various theories have been put forward, most, if not all, of which postulate that at the time that this drone-producing egg is passing down the oviduct on its way to the exterior the muscular sphincter at the mouth of the spermathecal duct remains tightly closed and thus prevents the escape of sperm into the oviduct. One of the first theories put forward, for example, supposes that when a queen lays an egg in a worker cell her abdomen is somewhat compressed by the narrowness of the cell and this results in the relaxation of the sphincter and release of sperm into the oviduct; whereas, when a queen is laying an egg in a drone cell with its greater diameter her abdomen is not compressed, the exit from the spermathecal duct remains closed and thus no sperm are released to fertilise the egg. The obvious difficulty with this theory is that it fails to explain how it is that a queen will on occasion lay fertilised, worker-eggs in cells that are only partially built and whose sides are far too short to exert any pressure whatsoever on her abdomen.

Another theory supposed that the queen receives some stimulus from worker and queen cells,* probably by touching them with her antennae, with which she examines each cell before laying in it, and that this resulted in a reflex opening of the mouth of her spermathecal duct and the release of sperm into the path of the descending egg. We now know, however, thanks to the elegant experiments of Dr. N. Koeniger, that the queen receives a stimulus that prevents her releasing sperm when she examines a drone cell, rather than receiving stimuli causing her to release them when examining worker and queen cells. Furthermore, it is apparently not mainly through her antennae that she receives this stimulus but through her front-legs which, together with her antennae and head, she thrusts into cells before laying in them. In this way, perhaps by using her front-legs as a pair of callipers to measure the diameter of the mouth of each cell, the queen is able to distinguish between drone cells on the one hand and worker and queen cells on the other. When Dr. Koeniger prevented a queen from putting her front-legs into cells she was examining before laying in them, by attaching little strips of Scotch tape to the tibia of both of her front-legs so that they stuck out like spurs, almost all the eggs she laid in drone cells were fertilised ones. It seems, therefore, that a queen receives a specific stimulus from a drone cell which in some way not yet understood usually results in inhibition of fertilisation of an egg that is laid in it. Sometimes, however, it is probable that "mistakes" occur and that fertilised eggs are laid in drone cells, either because a queen fails to receive the right—no sperm—message from a drone cell or, perhaps, because she has just moved from a worker cell in which she has laid a fertilised egg to a drone cell and there are still live sperm in her oviduct when she lays in the latter. What happens to the eggs? Do the workers remove them? Such behaviour seems possible, especially as Dr. J. Woyke has shown that worker bees can distinguish between fertilised and unfertilised eggs and on finding fertilised eggs in drone cells often remove

*Some people have suggested that queen honeybees do not lay eggs in queen cells and that the eggs that are found in them are conveyed thither by worker bees. That this is certainly not always the case is shown by the fact that various observers, including the author, have watched queens ovipositing in such cells.

and eat them, although they did not behave in this way in Dr. Koeniger's experiments.

We have seen that all female honeybees, that is to say workers and queens, are produced from fertilised eggs. Genetically and in every other respect their eggs appear to be identical, the only obvious difference being that those which produce workers are found in worker cells, and those few which give rise to queens occur in the very differently shaped queen cells. If we take a young larva, that is less than about three days old, from a worker cell and transfer it into an empty queen cell, the bees are able to rear a queen from it; similarly if we transfer a young larva from a queen cell into a worker cell the bees will, in many cases, rear a worker bee from it, but never a queen. The experiment of transferring a young larva from a worker cell into a queen cell or, rather, into an artificial queen cell, has been repeated success-fully literally thousands, probably millions, of times; as it forms the basis of the various methods by which breeders rear queen honeybees.

Many people have attempted to answer the difficult, and indeed still incompletely resolved, question of how bees can produce queens and workers from the same kind of eggs. In 1888 von Planta published the results of his analyses of the food of larval honeybees, which appear to show that during the first three days of larval life the young larva that is destined to become a queen is fed on precisely the same kind of food as that which is destined to become a worker, or, for that matter, a drone. However, his data appear to show that on the fourth day the food that is given by the bees to larvae that are destined to become workers is changed, pollen and nectar (or dilute honey) being added to it; whereas larvae in queen cells continue to be fed on the same kind of food throughout larval life. This special kind of milky-white food which von Planta believed to be fed to queen larvae throughout their lives and to worker larvae only for the first three days of larval life, became known as "royal jelly". For many years it was uncertain whether this food is regurgi-tated in a pre-digested form by the nurse bees or whether they secrete it in special glands. It is now known to be a protein-rich, highly nutritious food secreted by the hypopharyngeal (brood-food) and mandibular glands of the worker bees. Usually, but not invariably (as we shall see later) the protein in this food is obtained directly from the large quantity of pollen which is normally eaten by the worker bees soon after they become adults. Thus it has come to be believed that any female honeybee larva (that is to say any larva that is derived from a fertilised egg) that is fed exclusively on royal jelly always develops into a queen bee, whereas any such larva which is otherwise fed develops into a worker honeybee.

This belief, based very largely on von Planta's analyses of larval food, has persisted to the present day. Exactly how this difference is effected by feeding has been studied by workers, who have made new analyses and comparisons of samples of the food that worker bees place in the cells of both queen and worker larvae of different ages. These indicate that larval food does not have any fixed composition but is, indeed, very variable. Nevertheless, the larval food found in queen cells containing larvae of all ages is always whitish and contains very few, if any, pollen grains; whereas the food found in cells containing the older worker larvae is often distinctly yellow, green or even red, due to the colours of the many pollen grains contained therein. These facts, which have no doubt been observed by many beekeepers, have undoubtedly been considered as confirmatory evidence of the larval foodstuff alteration theory. However, several workers have observed that the complete absence of pollen seldom prevents the rearing of worker larvae successfully for a few days. It is clear, therefore, that pollen grains are not absolutely essential ingredients in the diet of a larval honeybee that is destined to become a worker.

A more recent theory to explain caste differentiation (differentiation between queen and worker honeybees) was put forward by Haydak, who suggested that the production of either a queen or a worker from a female larva is due primarily to differences in the quantities of the essential nutrients consumed. He pointed out that until about the third day of life all female honeybee larvae are fed with almost unlimited quantities of food; their cells are mass-provisioned and the larvae are able to eat as much as they like. After the third day the queen

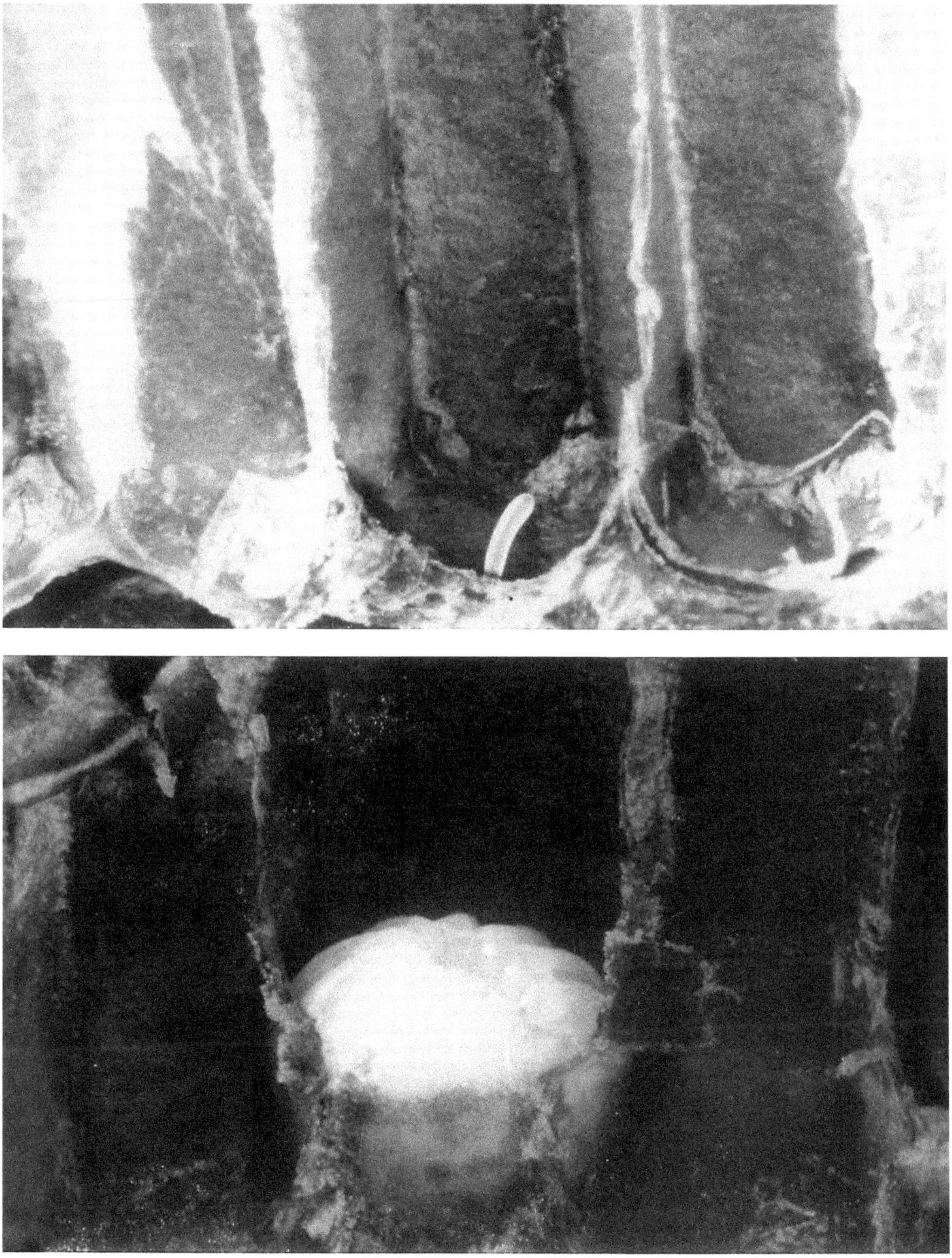

Plate 7.—Life History of Worker Honeybee (1)
above. Egg in worker cell (x 11).   *below.* Young worker curled up in cell (x 11).
(*See also Plate 8*)

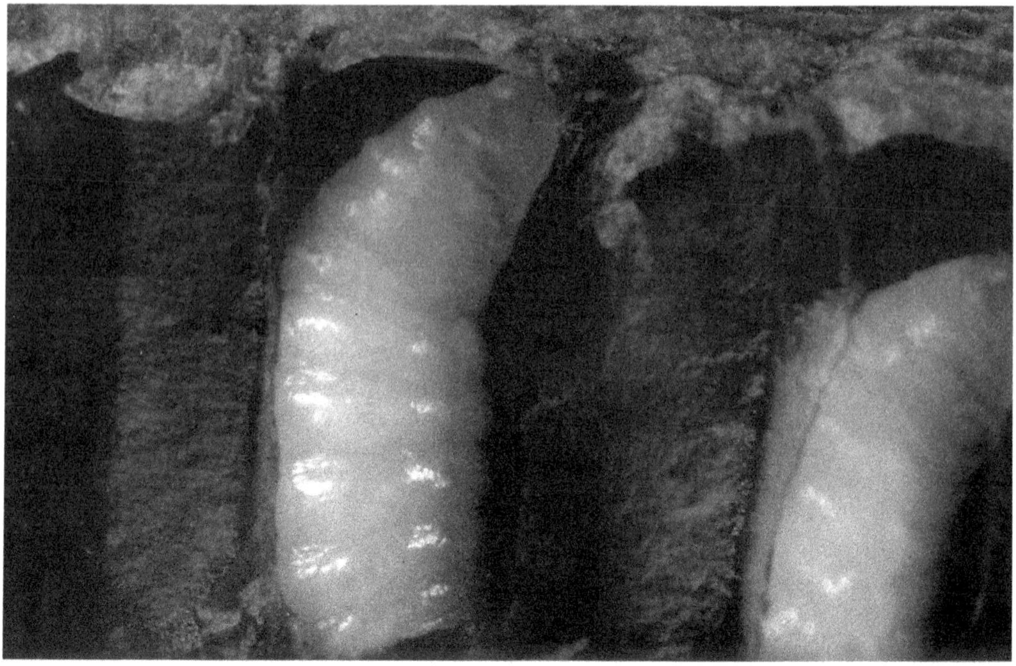

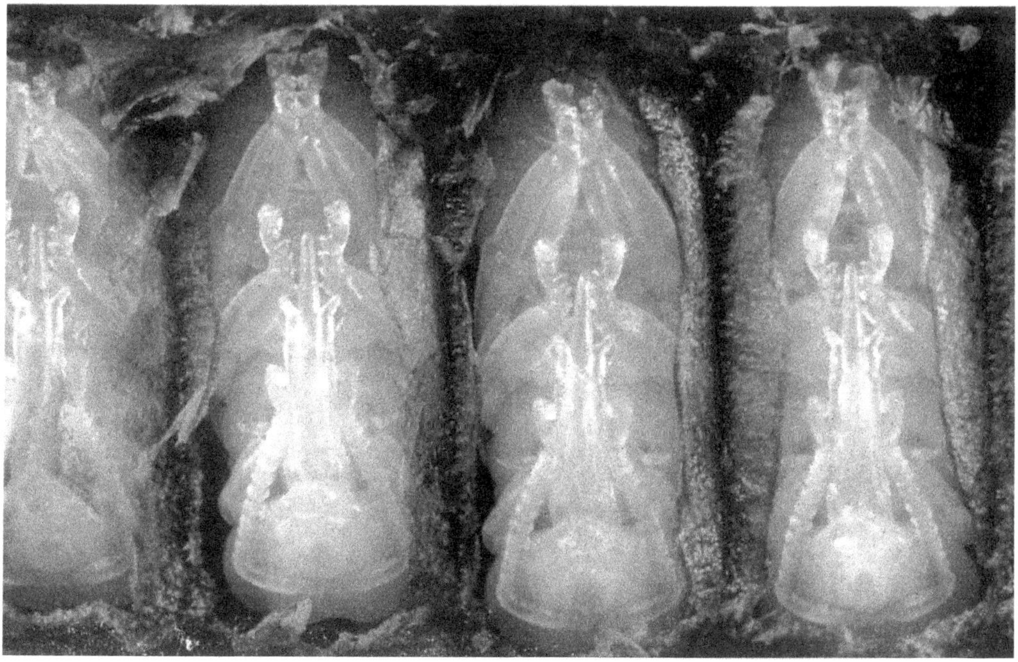

Plate 8.—LIFE HISTORY OF WORKER HONEYBEE (II)
above. Full-grown larva stretched out in cell (x 7·6).   *below.* Pupae in their cells
(x 6·0)

larvae continue to live amidst a superabundance of food, but the cells containing worker larvae cease to be mass-provisioned by the nurse bees. Progressive-provisioning now begins and continues throughout the remainder of the larval life, small quantities of food being fed to the larvae from time to time. It seems possible that any other form of feeding of an older worker larva which occupies so much of the limited space available in its cell would be impossible. The queen cell, on the other hand, is large enough to accommodate an abundance of food as well as a large larva. The result seems to be that, whereas queen and worker larvae grow at approximately the same rate during the first few days of larval life, after about the third day the queen larva, surrounded by an abundance, indeed a surplus, of rich food, continues to grow very rapidly. The larva in the worker cell, however, with its smaller diet grows much more slowly. The amount of surplus food fed to the queen larva is shown by the fact that even after the adult queen has emerged from her cell it is usual to find a large amount of semi-desiccated food still remaining behind in it, a thing never found in a worker cell (Pl. 12c, p. 83). Even after her cell has been sealed the queen larva continues to feed and to grow for some time before she pupates, whereas (once her cell has been sealed) the worker larva has nothing upon which to feed (Pl. 8a, p. 47) and actually loses weight between the sealing of the cell and the time when she pupates (Pl. 8b, p. 47). It might seem, therefore, that the quantity, rather than the quality, of the food eaten determines whether or not any particular female larva will develop into a queen or a worker. However, as already mentioned, there is often a difference in colour between the food in worker and queen cells containing larvae of about three or more days old, due to the presence of many pollen grains in the food in the worker cells and their almost complete absence in food in queen cells. It is also true that although various people, including my late colleague Mr. P. S. Milne, have succeeded in "hand-rearing" bees from very young larvae to the adult stage in the laboratory, very few queens have been produced.

This has happened despite the fact that the larvae were kept in specially made, large, wax cells in which plentiful supplies of larval food, from natural cells containing worker larvae, were constantly maintained. Such "hand-reared" larvae were certainly fed by the system of mass-provisioning and supplied with an abundance of food throughout larval life. Why was it then that none of them developed into queens? Similarly another of my colleagues, Dr. J. Simpson, has shown me larvae in worker cells taken from very populous colonies from which all but perhaps half-a-dozen larvae had been regularly removed, and such larvae were almost swimming in food but were not differentiating into queens.

Haydak approached this matter in another way. Instead of taking young larvae from worker cells and providing them with abundant food to see whether queens could be produced in this way, he took older larvae from queen cells that were either in the process of being sealed, or had just been sealed, and kept them away from further food. Thus he prevented them from continuing to feed upon the mass of food stored in their cells after the latter had been sealed. Most of the larvae which he treated in this way died as pupae, and seven definitely possessed worker rather than queen characteristics, whilst some of the others had characteristics that were intermediate between queens and workers. Nine living adults were produced, of which eight were normal queens and one an intermediate creature with the head, mandibles and sting of a typical worker, but the body, legs, etc. of a queen. Haydak had taken care to weigh the larvae when he removed them from their queen cells, and he was able to show that the average initial weight of those of the larvae which developed into queens was 14% higher than that of those which developed into the individuals with certain worker-like characteristics.

From the results of the experiments described above it would seem probable that partial starvation, from about the third day of larval life onwards, of those larvae that are destined to become workers—or, if one looks at it the other way round, excessive feeding of those larvae that are destined to become queens—plays an important part in the differentiation of the worker bee from the queen bee, both being capable of being produced from any young female larva. But, as we have seen, it does not appear to be quite the whole story, or surely those larvae which have been hand-reared and kept in cells in which a surplus of food has been constantly maintained around them should have produced a few adults which—even if they were not full queens—at least showed some signs of being intermediate between queens and workers.

The German worker von Rhein was probably the first person to rear bee larvae by hand in the laboratory. He fed young larvae that he had taken from worker cells with an abundance of "royal jelly" which he obtained from queen cells containing fairly old larvae. Instead of producing queens, he got worker bees with few, if any, queen-like characteristics. He therefore concluded that some special hormone-like growth-promoting substance is added by the bees to the food of those larvae that are destined to become queens, at an early stage in their development. (Larvae more that about three days old cannot, apparently, be converted into queens by the bees). He explained the fact that he had not succeeded in producing queens by feeding "royal jelly" taken from queen cells, by suggesting that this hormone-like substance is either fed directly by the nurse bees into the mouths of the queen larvae, or else is of such an unstable and fugitive nature than it had disappeared from the "royal jelly" which he had collected and stored for subsequent feeding to larvae. Here it should be mentioned that nurse bees have not, so far as I can discover, ever been seen to give food directly to any larva, but instead always place the food in the cell beside the larva. A larva in a queen cell comes in practice to rest upon a veritable bed of "royal jelly" (Pl. 12c, p. 83) so greatly in excess of her immediate requirements is the quantity of food that is fed to her.

Other workers, besides von Rhein, have suggested that a substance (or substances) is present in the food worker bees give to larvae that are destined to become queens which is not present in the food they give to worker larvae, and have supposed that this substance promotes differentiation between workers and queens. Indeed it appears from the work of Dr. H. Rembold and others that the "royal jelly" that worker bees feed to larvae that develop into queens is very similar to that fed to those that become workers except that the former has a rather labile differentiating substance (or substances), which is probably produced in the worker's mandibular glands, added to it. The labile nature of the differentiating substance, which has not yet been identified, probably explains why so many people have failed to rear queens in the laboratory by feeding female larvae on abundant food collected from queen cells and stored. When, however, Dr. S. C. Jay transferred larvae to freshly collected royal jelly three times a day about half of them developed into queens.

The fact remains, however, that, although it is possible to rear a newly emerged larva to maturity in the laboratory, nobody has yet been able to rear a queen from such a larva by adding any particular substance to its food. Until this, or some similar critical experiment, has been successfully completed, we shall not know for certain the answer to this intriguing problem. From the data that are at present available, it seems probable that the quantity of food eaten is important, but that quality of food also plays an indispensable part in the story. It seems likely that a certain minimum quantity of some special substance (or substances), which is present in the food fed to larvae in queen cells but not in that fed to larvae in worker cells, is necessary for full differentiation of the queen. If a female larva is fed with an adequate quantity of this food, from about the second day (perhaps a little earlier) of larval life onwards, it will develop into a queen. If, on the other hand, a larva is fed too small a quantity of this special food, perhaps because it was not given any until it was three days old or more, either a worker bee or (if the quantity was sufficient to cause some differentiation but insufficient to cause complete differentiation) a bee with characteristics that are intermediate between those of a queen and a worker is produced.

One other interesting fact should be remembered as it may possibly, though I think improbably, have some bearing on this problem of caste determination. It is that a typical queen cell hangs almost vertically downwards from the face or sides of the comb with its contained larva curled up in the horizontal plane and held on its bed of royal jelly by surface tension (Pl. 12c, p. 83), whereas the long axis of the worker cell lies almost horizontally and the larva lies in a different plane, an almost vertical one, than does her sister in the queen cell (Pl. 29a, p. 162).

Whatever the determining factor or factors may be, the fact is that until it is about three days old any female larva is capable of becoming a queen given the right treatment. Once the larva is more than about three days old this is no longer possible, and if a larva of this age is transferred to a queen cell, and the bees of a colony can be induced to accept it (quite a difficult thing to get them to do) a bee in which worker- rather than queen-characteristics preponderate is produced. It appears to be essential that differentiation should have begun before a larva is more than three days old if a queen-like adult is to be produced, and earlier than this if a really perfect queen is to result. Presumably the developing ovaries or other organs of a larva that is being reared in a queen cell produce and release a hormone (or hormones) into its blood which inhibits the development of pollen-baskets, straight sting, and other worker characteristics which are no longer possessed by the queens of present-day honeybees, but which were almost certainly possessed by their more primitive ancestors, just as the relatively primitive bumblebee queens possess them today.

We shall see later (Ch. 5, p. 52) what an important part the method of caste determination practised by honeybees plays in colony survival.

CHAPTER 5

ON THE SURVIVAL
OF COLONIES THAT HAVE LOST
THEIR QUEENS

> *"A colony of bees cannot subsist without a queen, as she alone produces their numerous posterity; and on this account their fidelity and attachment to their sovereign is admirable."*
>
> JOHN MILLS (1766). *An Essay on the Management of Bees*

NORMALLY a colony of honeybees possesses a single mated queen who alone provides the eggs from which all the worker and drone members of her colony are produced. It is obvious that it is a serious matter when a colony loses its queen through old age, disease or accident. Were it not for the fact that both workers and queens can be reared from any young female larvae (p. 45), such colonies would soon perish as their members died off and no young bees became available to take their places.

When a colony loses its queen the worker bees soon become aware of the fact (often within half-an-hour or less) and behave in a very characteristic manner. Some of them run around, either at the entrance of the hive or inside it, as if seeking their lost queen. If one opens a hive containing a colony which has recently lost its queen one frequently hears a strange sound, which is usually called "roaring" today but was called "moaning" by many of the old beekeepers. This appears to be caused by a large number of bees standing more or less still on the combs, and rapidly fanning their wings so that a current of air is driven over certain scent-producing glands which they simultaneously expose. This scent smells somewhat "flower-like" or, perhaps, as another observer has described it, rather like slightly musty hay, and, as we shall see later, such bee-produced scent is very attractive to other bees.

There is little doubt that the stimulus which resulted in the bees dispersing scent in this way, and producing the "queenless roaring sound", was the opening of their hive. As a rule no roaring occurs in a queenless colony if their hive is left undisturbed. When one opens the hive of a queenless colony, one can often see that a small group of bees begins to fan in this way and that the behaviour of these bees appears to excite other members of the colony to behave in the same way. A "snowball effect" sometimes takes place, until almost all of the bees are fanning and the roaring is intense. Since scent-dispersal by a few bees of a colony serves to attract other members of the colony, and often to cause them to begin distributing scent also, it seems reasonable to suppose that the few bees which first began to disperse scent when their hive was opened, stimulated more and more bees to behave in the same way as soon as they smelt the perfume produced by the fanning bees. As we shall see in a minute, these bees behaved as though they had found their queen.

Scent communication amongst a group of bees is often built up in this way. For example, if a number of queenless worker bees are trying to escape from a cage and we allow one of them to find an exit from the cage, she will often stand near the exit and begin to fan with exposed scent gland. The scented current of air stimulates other bees along its line of flow to behave in the same way also, so that a whole column of fanning bees with exposed scent glands often develops from the exit well into the cage.

If, then, we take the queen away temporarily from a colony, sooner or later the bees will almost certainly begin to behave in this characteristically queenless way, and some of them will begin to run about excitedly around the entrance of the hive as though seeking their lost queen. If we now return the queen to her hive, releasing her amongst the group of bees at the entrance, so that we disturb the colony as little as possible and do not set its members "roaring" by opening their hive, we find that the behaviour of the bees nearest to the returned queen changes immediately. They stop running about and some of them begin to examine, lick, groom and feed her, whilst others expose their scent glands and drive currents of air over them with swiftly fanning wings. Within a short time bees within the hive often begin to disperse scent as well. Soon the queen enters the hive and the excitement gradually subsides until the colony has returned to normal once more, usually within twenty minutes or half-an-hour of the return of the queen. The hive can then be opened without causing the characteristic "queenless roaring".

If, instead of returning the queen to her hive, we place an empty wire-gauze cage, in which the queen has been confined for some time and from which she has just been removed, at the entrance to the hive, the bees often behave in very much the same sort of way as they do when we actually return their queen. After exploring the empty cage for some time, however, the bees seem to realise that their queen is not in it and the typical queenless behaviour is resumed. It is clear that it can only be something, perhaps an odour, left by the queen on the cage walls which temporarily satisfies the bees and causes them to behave as they do towards an empty cage.

One is thus tempted to suppose that bees habitually recognise the presence of their queen by her scent and, conversely, realise when she is absent because of lack of her scent. Unfortunately, however, this simple theory does not explain everything, and the manner in which the bees of a colony do recognise the presence of their queen is so interesting, and so important a factor in maintaining the social cohesion of the honeybee colony, that it will be discussed separately (Chapter 8, p. 97). It is sufficient at the moment to realise that the bees of every colony can very definitely and quickly appreciate whether their queen is present or not.

The most important (biologically speaking), and the most definite, of all the signs of queenlessness which bees exhibit has not yet been mentioned. We have already seen that it is possible for worker bees to rear a queen either from an egg or larva found in one of the specially shaped queen cells, or from an egg or young larva found in a worker cell (p. 45). So strong is the urge of the bees of most colonies to obtain another queen to replace the one that they have lost that within an hour or two of appreciating their loss they begin to rear a new one from a young larva in a worker cell. Usually they attempt to rear several queens simultaneously in this way. The bees enlarge worker cells containing young larvae and reshape them into the form of queen cells, at the same time giving the young larvae a great deal more food than they would normally receive. Fresh new wax becomes visible around the mouths of these worker cells. Such "emergency" queen cells can be distinguished both from "supersedure" and "swarm" queen cells even after the bees have completed them, as their genesis from a worker cell remains apparent, especially in longitudinal section. (Pl. 13, p. 86). "Supersedure" and "swarm" queen cells first make their appearance as special cups, often called "queen cups" or "cell cups", which the worker bees build usually at the sides of the combs, or in other positions in the brood nest where there is sufficient room for development of the completed queen cells (Pl. 12a, p. 83).

In spite of some opinions to the contrary, the queen herself will lay eggs in "queen cups" as I myself have observed in observation hives. Thus both "supersedure" and "swarm" cells start as queen cells and have, unlike "emergency" cells, no connections with worker cells (c.f. Pl. 12, p. 83 and Pl. 13, p. 86). "Supersedure" cells are built when the bees apparently "wish" to replace a queen that is failing to supply enough "queen substance" to satisfy them (see Chapter 8). Such cells cannot individually be distinguished from swarm cells. Most colonies that are about to swarm build numbers of queen cells* whereas a colony that is trying to supersede its queen seldom produces more than two or three queen cells, often only one.

By making use of the fact that the larva from any female egg (i.e. fertilised egg laid in a queen or worker cell) can be caused to develop either into a worker bee or into a queen bee by means of differential feeding, worker honeybees are thus enabled to produce a new queen for themselves when they are unfortunate enough to lose their old one. The inherent disadvantage of each colony only possessing a single queen can be, and frequently is,

*Colonies of some races often build several dozen at a time, but those of the races commonly kept in Britain usually only produce about a dozen.

overcome in this way, provided that a female larva of less than about three days old is available to it.

Now it sometimes happens that a colony loses its queen at a time when it does not possess any eggs or young female larvae. Are such colonies doomed to extinction as the worker bees gradually die off? The answer is that in the vast majority of cases they do dwindle away and die out; but in a small minority of cases the bees of such "hopeless" colonies somehow acquire a new queen for themselves and thus survive. This phenomenon, which is certainly very rare amongst colonies of those races of bees most commonly kept in Britain, has been called "the mystery of the unexplained queen". In attempts to explain the sudden appearance of such queens in colonies known to possess no eggs or larvae and, therefore, no apparent means of rearing a queen, it has been suggested that the worker bees have stolen a fertilised egg or young larva from another colony. This appears to be a very unlikely happening, as does the suggestion that a worker honeybee will exceptionally mate with a drone and produce a few fertilised eggs from one of which a queen is raised. Neither of these notions appears to be supported by any evidence.

Many people have seen worker bees with eggs between their mandibles; but no-one appears to have seen workers putting such eggs into cells, and it is possible that they were about to eat them. However, it is possible that workers sometimes transfer fertilised female-producing eggs laid by a queen to queen cells, because eggs, presumed to be of this kind, have been found in queen cells in parts of hives from which queens were excluded. Similarly, I have found single eggs, properly positioned in queen cells and therefore unlikely to have been laid by workers, in colonies whose queens were experimentally tethered and, although producing eggs, were not only prevented from reaching the queen cells but also unable to lay in cells.

It is a well known fact that in many cases if a colony of honeybees loses its queen and, for some reason or other, fails to raise an emergency queen, thus becoming "hopelessly" queenless, eggs begin to appear after a time in worker cells and even in queen cell cups. In such cases the individual cells often contain not one but several eggs scattered about in a rather haphazard manner (Pl. 16, p. 95), very different from the orderly one-egg-per-cell arrangement produced by normal queens. These eggs have in fact been laid by workers, such bees being known as "laying workers". Since the worker honeybee is physically unable to mate and to become fecundated, it follows that any eggs which she lays must be unfertilised and give rise to drones. As a result one often sees the characteristically domed cappings which bees use to seal drone cells applied to worker cells in a colony in which laying workers are active; and sometimes undersized, but fertile, drones that have been reared in worker cells are present in numbers. Most of the eggs, or the larvae which emerge from worker eggs are, however, thrown out of the hive by the worker bees. One sometimes finds several young larvae in a single cell (Pl. 16, p. 95) but at most only one survives.

Laying workers can be divided into two categories—potential laying workers whose ovaries have become more highly developed than usual, and functional laying workers whose ovaries are active and which are, or very soon will be, laying eggs. As many as eighty per cent, or even more, of the bees of a colony may become laying workers. There is evidence that this unusual development of their ovaries is favoured by diversion of nutriment from their brood-food glands—in the absence of an outlet for the secretions of these glands in the shape of enough larvae to be fed. Thus anything which upsets the balance between the number of larvae requiring to be fed and the number of bees with functional brood-food glands prepared to feed them—such as a sudden check in egg-laying by the queen—tends to result in the production of potential laying workers. However, such potential laying workers seldom appear to become functional unless their queen—and thus inhibitory "queen substance" (Ch. 8, p. 97)—is actually removed.

Now Mackensen has found in his work with two distinct races of bees (the Italian and Caucasian races) that a very small percentage (not more than 1 per cent) of the unfertilised eggs laid by unmated, virgin queens give rise to females. The majority, of course, give rise to drones. He found that both workers and queens were produced from such parthenogenetic (unfertilised) eggs, and he actually succeeded in obtaining two generations of parthenogenetic queens, these queens appearing to be perfectly normal individuals with the female number of chromosomes. The number of chromosomes in the unfertilised eggs from which these females were produced somehow became doubled in number before their embryos commenced to develop. Similar females produced from unfertilised eggs have been reported in Syrian and Tunisian races of honeybees, as well as in some from S. Africa, so that it appears that this unexpected phenomenon can occur in a number of different races of honeybees, although it is only rarely that it becomes apparent. *

During 1943 and 1944 when studying the behaviour of laying workers in colonies of Italian bees, the author had under observation at one time or another about twenty queenless colonies in which functional laying workers were present. Eggs and sometimes larvae appeared in queen cell cups (Pl. 16b, p. 95) in most of these colonies; but only in one colony were any sealed queen cells seen. This colony had been queenless for about five weeks and was at least half-a-mile away from any other colony. From one of the three sealed queen cells found in this colony a virgin queen—one of the unexplained variety—emerged, the other cells together with their occupants being destroyed by the bees. This virgin queen mated successfully and produced normal worker brood, thus saving her colony from inevitable extinction.

It appears highly probable, therefore, that those queens that have been found by beekeepers unexpectedly in hopelessly queenless colonies have arisen from female larvae produced exceptionally from unfertilised eggs laid by laying workers. Thus, even when the bees of a colony fail to raise emergency queen cells from worker cells containing young female larvae, there is still a chance, albeit a small one, that they will succeed in rearing a queen and thus ensure the continued existence of their colony.

---

*Dr. R. H. Anderson of S. Africa has now shown that in the case of the so-called Cape Bee (*Apis mellifera capensis*) many eggs laid by workers produce female larvae which develop into workers. Indeed, queenless colonies have been known to persist for months. Apparently the Cape Bee is very loath to rear new queens by means of emergency queen cells.

CHAPTER 6

THE LIFE OF A QUEEN BEE
MATING, EGG-LAYING BEHAVIOUR, ETC.

*"The Queen bee is a fair and stately creature, longer by the half,
and much bigger than a common Honeybee, yet not so big as a
Drone, but somewhat longer."*
JOHN GEDDE (1721).
The English Apiary or the Compleat Bee-Master

THE QUEEN is the one indispensable member of every honeybee
colony. Without a mated queen a colony dwindles away as
its members die of old age and no others come along to take their
places. Only a mated queen can produce the large number of
fertilised eggs necessary for replacing the worker members of the
colony (perhaps as many as 200,000 a year in the case of a prosperous colony) as well as a few hundred unfertilised eggs from
which drones will come.

Not only is a mated queen essential because of her reproductive function; but she also provides the cohesive force which keeps
the thousands of worker members of the colony together as a
social unit. This she does by satisfying their demands for "queen
substance"—*i.e.* a substance secreted by the queen—and, possibly,
other materials. This social, cohesive, function of the queen is
extremely important in the maintenance of what has been aptly
called "colony morale" and we shall discuss it when we consider
the whole question of colony organisation (Chapter 9, p. 111).

We have seen how the workers of many colonies of bees are
able to rear so-called "supersedure" queens to replace those which
are failing through age, disease or injury. Such supersedure
queens are reared in normal queen cells in which the failing

queens themselves lay eggs. Often a supersedure queen will live quite happily for some months with her mother whom she is destined to replace, the two queens together providing the eggs necessary for the maintenance of their colony (Pl. 18, p. 99). We have also seen that, even in the event of the sudden and unexpected loss of their mated queen, the bees of a colony can rear a queen to replace her, provided that there is at least one sufficiently young worker larva present in the hive, whose normal worker cell they can convert into an emergency queen cell and flood with brood-food for the developing larva (Pl. 13, p. 86).

Let us, then, consider the future of a virgin queen, produced either from a supersedure, or from an emergency, queen cell. First, having cut a slit three-quarters of the way round her cell just beneath the capping with her mandibles, she pushes open the now neatly hinged capping and emerges amongst the worker bees that have reared her.

One might perhaps expect that the worker bees would pay a good deal of attention to the virgin queen after devoting so much energy and care to her production, but they generally appear to pay relatively little attention to her. If, however, she is removed from the hive for a short time, the bees soon begin to exhibit obvious signs of queenlessness and if she is returned to them they lick her body, feed her, and indeed pay great attention to her. It is clear, therefore, although it is not generally obvious, that the worker bees do, in fact, treat their virgin queen very differently from another worker bee. On her part, the virgin queen helps herself to honey from within the hive, but apparently eats no pollen. She occasionally solicits food from worker bees and appears to behave very much like any ordinary worker except for the fact she is not known to undertake any hive duties. Instead, a short time after she has emerged from her cell she starts to wander around the brood combs, seldom straying far from the brood area unless disturbed by somebody opening her hive.

It is probable that the worker bees of a colony will have built two or three supersedure cells or, perhaps, as many as a dozen or more emergency queen cells, distributed on one or more combs. Usually these cells contain larvae which differ slightly in age, so that one of them develops into an adult virgin queen and emerges from her cell a short time before any of the others. However, should two virgin queens emerge almost simultaneously from their cells and meet one another whilst wandering around in the hive, a bitter fight takes place between them. Each grapples with the other and attempts to sting her, until one of them succeeds in killing the other by piercing her body with the sting—as a rule through one of the relatively soft membranes which separate the hard chitinous plates with which the segments are protected, and give flexibility to the body. The worker bees do not appear to take much notice of such fights but soon carry out the corpse of the vanquished from their hive, just as they dispose of the corpse of any worker bee which dies. However, as already mentioned, it usually happens that one of the virgin queens emerges before her sisters are ready to do so. If, during her wanderings around the brood combs, this queen comes across a sealed queen cell, she almost always exhibits great excitement, climbs on to it, and attempts to tear a hole in it with her mandibles. Curiously enough she seldom seems to pay much attention to any unsealed queen cells she comes across.

Those queens that the author has watched attacking the cells of their rivals often started to tear holes with their mandibles close to the tips of the cells, but sooner or later transferred their attention nearer to the bases. Here, after what appeared to be a great deal of effort, they chewed small holes and then wandered off, perhaps to find another queen cell and to repeat the process. It has been reported that when she has found a queen cell containing a potential rival and has torn a hole in it in this way, the queen turns round and, inserting the tip of her abdomen into the hole in the cell, proceeds to sting the helpless occupant to death.

I must admit that I have never seen this happen, possibly because those queen cells which I have seen attacked have all contained full-grown larvae or pupae rather than adults. Anyway the end results were usually much the same, for when the workers found these holes in the cells they enlarged them, pulled out the occupants piece by piece, and eventually tore the cells right down. However, on one occasion some workers found a hole that a queen had made near the base of a queen cell and carefully repaired it instead of destroying the cell and its occupant. Once again the queen chewed a hole in this cell and once more the workers repaired it. When the queen made a hole in it for the third time the bees did destroy it.

The first virgin to emerge often makes a series of curious shrill sounds—z-e-e-p, z-e-e-p, zeep, zeep—which is aptly called "piping". It has been stated that such sounds stimulate any other virgin queen in the hive to reply in the same way, though she may still be in her cell. Furthermore, it is said that such piping excites and assists the rivals to find one another and fight it out.

Exactly how the piping sounds are produced was for long a matter of controversy. Some people thought they were made by the forcible expulsion of air through the queen's spiracles (openings of the breathing system to the exterior), while others believed that they were, as has now been demonstrated by Dr. J. Simpson, produced by operation of the flight motor without the wings being spread while the queen presses her body down on the comb. In this way the sounds are transmitted through the substrate to other bees, both workers and queens, the queens often responding by piping in reply. They will even do so to suitably transmitted tape recordings of queens piping.

The sound emitted by a queen that is still imprisoned in her cell sounds deeper than that of a free queen and has, I believe, given rise to the suggestion that whereas a young queen pipes shrilly an old queen "qualks" hoarsely, or perhaps it is the other way round.

The worker bees appear to pay little attention to the piping of the queens and may very possibly not even hear them—nobody has yet been able to demonstrate that worker bees can hear sounds transmitted through the air, though they can appreciate notes conveyed to them by contact with a vibrating surface. This was shown by J. B. Free and the author when we found that a queenless group of worker bees was attracted towards a tin containing a mechanical vibrator and settled on its surface (p. 118).

By the time she is a day or two old a virgin queen will usually have disposed of any rivals or potential rivals (we shall discuss certain exceptions to this rule when we consider the question of swarming) and, provided that the weather is suitable, she begins to take a number of short flights from her hive. These flights only last for a minute or two at first, but are gradually extended until they may last for ten or fifteen minutes or even more. During these flights the young queen learns the position of her hive relative to neighbouring objects and the topography of the countryside for some distance around, perhaps within a radius of half-a-mile—we do not have enough data to reach a precise estimate of the area around her hive known by the queen.

In my experience in England, virgin queens usually mate about the tenth day of adult life, but it is clear from the literature that they may mate when only four or five days old, and I myself have known of queens that did not mate until they were sixteen days old.

It used to be believed that each queen mates only once in her life, but we now know that most queens mate with several drones, often in quick succession on the same nuptial flight, before laying any eggs. Indeed, it seems clear that the majority of queens mate with about five drones before starting to lay. Furthermore, although there are a few reports of old queens leaving their hives on short flights, there does not appear to be any reliable evidence that queens that have once laid eggs ever mate again.

The sperm received from these drones remain alive, stored away in the queen's sperm-reservoir, or spermatheca, and serve to fertilise the several hundred thousand worker eggs which she may lay during the course of her life. Sometimes an old queen's supply of sperm becomes almost, or quite, exhausted and she then becomes a "drone-layer", being only able to produce unfertilised, and therefore drone-producing, eggs.

Where does the queen mate? Certainly somewhere outside the hive. We now know that the queen and drone meet in the air rather than on the ground and actually mate while flying. A number of observers have seen the comet-like formation of drones chasing a queen, darting and swerving hither and thither —sometimes 100 or more feet from the ground, sometimes only a few feet above it. This makes it very difficult to see exactly what is taking place, but it is generally believed that the "comet" is formed of drones with the queen at its head. On several occasions when I have witnessed this phenomenon the bees suddenly swooped down to the ground and by the time I had reached them all that I could find were a few rather dazed-looking drones crawling about in the grass.

Sometimes a queen is absent from her hive on her nuptial flight for as long as half-an-hour; indeed, even longer absences have been reported. How far away from her hive will a queen fly in order to find a mate? It has been stated by a number of writers that worker honeybees can fly at what appear to be quite fantastic speeds; some of the most extravagant estimates exceed fifty miles an hour! Actually the worker honeybee seldom flies at an average speed of more than twelve or fifteen miles an hour. She can move more quickly than this in short bursts of speed and may then reach twenty or even twenty-five miles an hour with the help of a following wind. However, such relatively high speeds are rare, and bees will seldom continue to forage for long when the speed of the wind exceeds fourteen or fifteen miles an hour. It is possible that the virgin queen honeybee can fly a little faster than a worker, but there do not appear to be any satisfactory experimental or observational data on which one may base a conclusion on this point. If we assume that, like the worker, the virgin queen can fly at an average speed of about fourteen miles an hour in still air, and, further, that she flies continuously all the time that she is away from the hive, she cannot fly more than about seven miles even should she be away for as long as thirty minutes. This means, of course, that the average queen probably mates within a radius of a mile or two from her hive.

Most virgin queens make their orientation flights, and their mating flights, sometime between noon and three or four o'clock in the afternoon (G.M.T.), i.e. at those times when drones are most active on the wing. In contrast to statements made by some beekeepers and, presumably, based on their own observations, all the virgin queens which I have seen leaving their hives on what subsequently proved to be their mating flights, have (with two exceptions) done so unaccompanied by either drones or workers. Indeed, the drones of a colony appear to pay no atten-tion even to a nubile queen in their own colony. In each of the two exceptions which I have mentioned a small group of several hundred worker bees, together with a number of drones—a so-called "mating swarm"—left the hive with the nubile queen and after flying around in the air for a few minutes settled in a swarm-cluster on a branch of a nearby tree. In one case the swarm-cluster was captured and returned to its hive and, on examination, the queen was found to have mated very recently and still to be carrying the "mating sign" (this will be described later on, p. 67). In the other case the queen was seen running on the surface of the swarm-cluster also bearing the "mating sign" and this swarm soon took to the air and returned to its own hive once more, about twenty minutes after it had first left it with a virgin queen. In one of these cases the queen was seen to leave the hive and, as usually happens with a swarm, came out after a good number of her workers had already left the hive. I suggest that whilst such "mating swarms" should not be confused with normal "colony reproductive swarms" it may well be the case that the latter evolved from a type of "mating swarm" (p. 154) (namely that in which only a proportion of the members of the parent colony leave the hive with the virgin queen, the rest remaining behind with another one, and in which the swarm seeks a new home after the queen has mated). Such "mating swarms" often occur with A. indica when a queen leaves a large colony to mate.

During the last fifty years several well known and reliable beekeepers have suggested that drone honeybees do not fly about at random seeking queens with whom to mate, but instead congre-gate together in certain definite places to which nubile virgin queens also resort. Several such restricted areas to which drones and queens resort year after year were discovered by Dr. F. Ruttner in Bavaria some years ago and others have since been discovered elsewhere, chiefly in mountainous places. In Britian and other countries few, if any, areas to which large numbers of drones regularly resort have been found. Perhaps in Bavaria few suitable mating places exist, so that those that do become very apparent, whereas in Britain and elsewhere there are so many that they do not become apparent.

Until recently few people had seen queen and drone honeybees mating. However, the discovery by Dr. N. E. Gary that free-flying drones are attracted by, and will often try to mate with, a queen when she is suspended by a thread twenty or thirty feet above the ground has enabled many observations to be made.

A virgin queen becomes nubile when she is about 7-10 days old and, if the weather is favourable, will leave her hive in search of drones with whom to mate. Not until she is flying at a height of about fifteen feet or more above the ground do drones appear to pay the slightest attention to her; even those leaving the hive in search of mates at the same time as the queen is leaving on her nuptial flight ignore her. Once a queen is flying at the right height, however, the odour of an acid, 9-oxodecenoic acid, which is secreted in her madibular glands and becomes spread over her body (see p. 107), attracts drones who are flying to leeward of her and not more than about 50 yards away. On perceiving the queen's scent, the drones fly up-wind on the lookout for her. Often, a number of drones see her more or less simultaneously and dash after her, so forming the characteristic "comet of drones" that have been described from time to time in the literature.

Soon one of the drones catches up with the queen and, having thoroughly examined and smelled her, seizes her from behind and above with his front and middle pairs of legs, curves the tip of his abdomen downwards and, by a violent contraction of his abdomen, everts his genitalia from within his body into the queen's open vagina where they become lodged. Eversion of his genitalia paralyses the drone causing him to release his hold of the queen who is still flying along and, still attached to her by his genitalia, to fall backwards away from her. A second

Plate 9 (opposite). THE GIANT HONEYBEE (*Apis dorsata*)
In Southern Asia colonies of the Giant honeybee build their single-comb nests in the open, attaching them to the undersides of limbs of tall trees, beneath bridges and the eaves of tall buildings, etc.
Above. Colony on comb attached to underside of overhanging rock. This comb was just over three feet long.
Below. Worker collecting nectar and pollen from a wild sunflower (x 2.7).

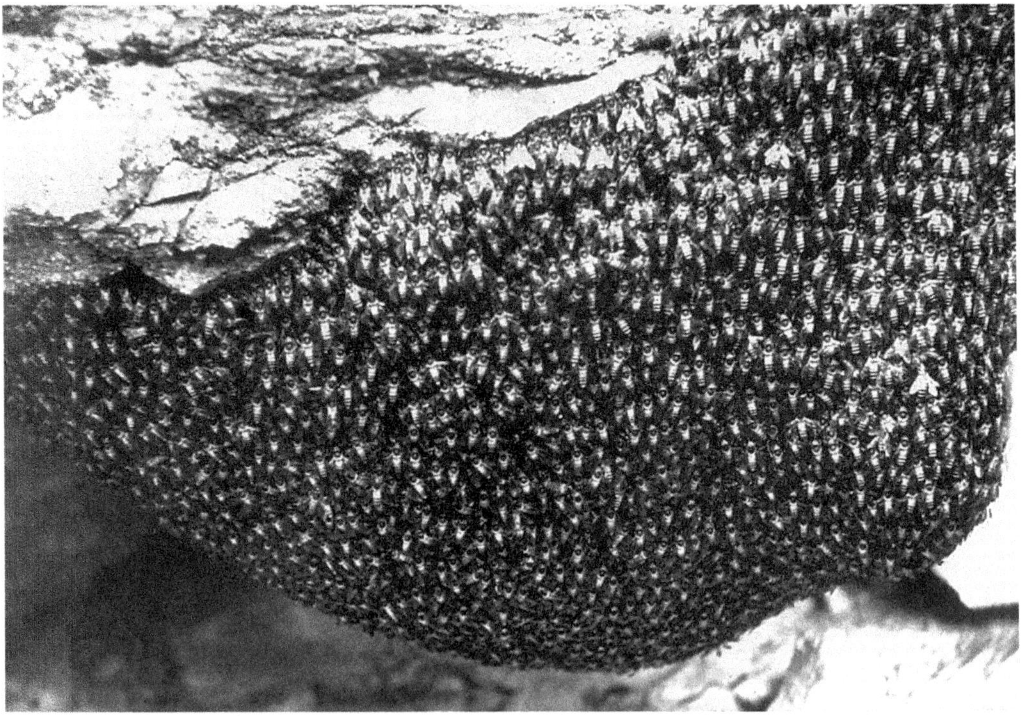

Plate 9

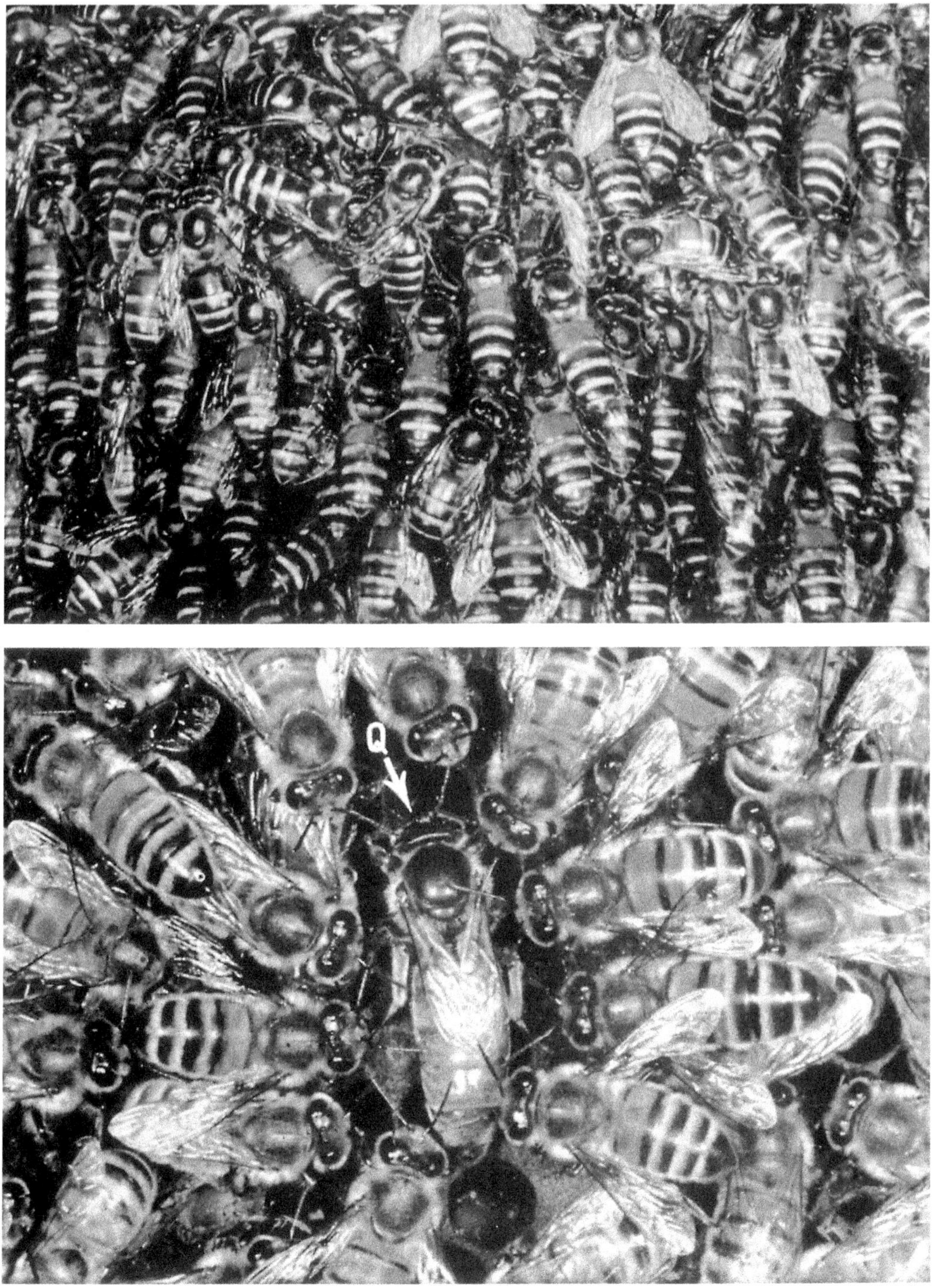

Plate 10.—Comparison of the sizes of the Little Honeybee (*Apis florea*) and the Western Honeybee (*Apis mellifera*)
above. Part of a colony of the Little Honeybee (x 2·7). *below.* Part of a colony of the Western Honeybee (x 2·7)

or two later a sharp explosive snap can sometimes be heard. It is due to compression, as he falls backwards from the queen, of air in the drone's genitalia and is sufficient to rupture them and release the queen who flies on with part of the drone's genitalia projecting from her vagina—the so-called "mating sign". The dead drone falls to the ground, another seizes the queen and in mating with her somehow displaces the remains of the genitalia of his predecessor.

Most queens mate with several drones either on the same or subsequent mating flights. Eventually, having been released from the last drone, the queen returns to her own colony, often with the "mating sign" protruding from her vagina. However, it soon shrivels up and drops off.

We have already mentioned how the worker bees of a colony appear to pay little special attention to any virgin queen they may have amongst them except on her return after being taken away for half-an-hour or so. But, when the queen returns from her mating flight, even though she may only have been away for a very few minutes, indeed too short a time for her absence to be felt (Chapter 8, p. 97), the workers of her colony pay rather more attention to her. They may catch hold of the remains of the drone's genitalia with their mandibles and tug at them. Within a day or two their attitude towards their newly mated queen undergoes a profound change. Some of the younger bees who happen to be nearest to the queen, in any part of the hive in which she may chance to wander, turn and face her and continually touch her body with their antennae and periodically lick it. When she solicits food from them with outstretched tongue the workers feed her (Pl. 20b, p. 115) on what I believe to be "brood-food", a proteinaceous and highly nutritious substance, secreted by their brood-food (hypopharyngeal) glands. When feeding the queen in this way a worker behaves slightly differently from when feeding another worker. When feeding the latter she opens her mandibles and moves her folded tongue slightly downwards and forwards from its position of rest beneath her head, at the same time regurgitating a drop of food which comes to rest between her mandibles and on the back of her tongue. (Pl. 21, p. 116). Such a drop of regurgitated food, into which the receiving bee sticks her tongue, is almost certainly mainly nectar or honey and not pure brood-food. According to the Russian observers Taranov and Ivanova, when a worker feeds her queen, she holds the drop of food between her mandibles as before, but she does not move her tongue downwards and forwards.

This difference leads me to suppose that forward and downward movement of the tongue is a necessary part of the regurgitation of nectar from the honey-stomach of a bee, but that it is unnecessary when bringing brood-food to the exterior from the collecting ducts of the hypopharyngeal glands preparatory to feeding a queen.

Soon after mating a queen's ovaries begin to enlarge and their growth is stimulated by the brood-food she receives. The size of her body increases enormously. Three or four days after mating she is ready to begin egg laying, and this she starts to do in any cells in the brood nest that are suitable. These are the cells in any comb lying within the "brood area"—an area where the temperature is constantly around 90°F or more—and have been prepared by the worker bees. Such preparation may involve removal of some old larval skins, the tamping down of others, and the tidying up and polishing of the cell walls, the result being cells with a clean, shiny, highly polished interior. It does not matter what kind of cell a mated queen comes across; provided it is within the "brood area" and is clean, she will lay in it. When laying her eggs the queen wanders around a comb in the "brood area", laying in cell after cell. She does not move in concentric circles as has been so often stated but wanders at random, crossing and recrossing her tracks and looking into each cell in turn. If she finds a cell that does not already contain an egg and is otherwise unoccupied she lays an egg in it. During the winter and early spring the "brood area" is confined to the middle of the winter cluster of bees and usually consists only of the centres of a few combs. Most combs contain only worker cells in their centres, any drone cells present tending to be nearer to the edges of the combs, so that as a rule only worker cells are included in the "brood area" of the colony in spring. Later, as the colony increases in size and the mean daily temperature rises, the "brood area" is extended until it includes whole combs, and thus the drone cells tend to become usable and the first drones of the year are reared. Strong colonies tend to include whole combs in their "brood areas" earlier in the year than do weaker ones and, as a result, they tend to produce drones earlier in the season. Occasionally, of course, a few drone cells occur in the centre of a comb, in which case drones may be produced exceptionally early. But often the bees refuse to rear drones in the early part of the season and remove and eat the eggs. Furthermore, sometimes when complete combs consisting only of drone cells are placed in a colony's "brood area" early in the season, the bees tear them down and rebuild with worker cells. It appears that, provided they have got laying queens, colonies of bees will seldom tolerate the presence of drones in their hives at those times of the year when they cannot collect much nectar in the field. Thus, just as bees will sometimes replace drone cells with worker ones in the early part of the year, they will also drag larval and pupal drones from their cells, and cast them out of the hive should a period of unfavourable foraging weather occur in the middle of the active season. Again, in the autumn, the workers of all queen-right colonies refuse either to feed their drones or to allow them access to the stores of honey. Soon the poor, starving drones drop to the bottom of the hive and are pulled outside by their sisters and left to die (Pl. 17, p. 98). I do not think that the workers ever sting the drones to death unless the latter are strong enough to attempt to escape by running. In this case, as we shall see when we discuss the question of hive defence (Chapter 12, p. 150), the stinging reaction of the worker bees is probably evoked and some of the drones do get stung. Normally, however, they are just dragged outside and left to die.

During the summer months a drone usually lives for about four or five weeks, becoming sexually mature when about ten days or a fortnight old. Drones never visit flowers in search of food, unlike males of other kinds of bees, but obtain all the food they require either by soliciting it from the worker members of their colony or by helping themselves from the stores within the hive. Whilst in the hive the drones congregate together in the warmest part which, of course, often means that area in which brood rearing is going on actively. Thus, quite fortuitously, they probably help to incubate the brood to some extent. The drones fly only on fine days, and usually only between noon and about five o'clock in the afternoon. A normal, well kept, colony will contain a few hundred drones during the summer but, as each queen is inseminated early in her life of three or four years (some queens have been known to live for as long as five or six years), the chance of any particular drone finding a mate is exceedingly small. On the other hand, with so many drones about, the chance of any particular queen failing to find a mate is also very small: thus the perpetuation of the species is ensured.

In the late spring and early summer the bees either build new, or renovate old, queen cell cups (Pl. 12a, p. 83). These are usually situated on the edges of the combs, or perhaps hang from the tops of any holes which may occur in the combs; indeed, in any positions which will allow room enough for the subsequent development of a complete queen cell. Many queen cups are never used at all; indeed the great majority are not. Unfortunately we do not yet know the stimulus which causes the bees to build these cups. Probably it is not the swarming impulse, as the cups appear in every colony every year whether or not swarming takes place. Some of the queen cell cups become included within the "brood area", in which case, should she happen to find them, the queen will often lay in them. If a queen lays an egg in a queen cell cup at a time when her colony is neither preparing to swarm nor to supersede its queen with a younger one, the worker bees sooner or later remove the egg, or the young larva which comes from it should it remain long enough to hatch.

It has often been suggested by beekeepers, and it is widely believed, that the worker bees of a colony actively and deliberately control the rate of egg laying of their queen as the occasion demands. It has been suggested that they stimulate her egg production by feeding her lavishly and, conversely, cause a decrease in her egg production by refraining from feeding her so lavishly. I believe that this notion is quite incorrect. Let us follow what I think is the course of events throughout the season. A colony builds up its strength, slowly at first and then at an increasing rate, until it reaches its full development, usually in Southern England towards the end of May or early June. This maximum strength is maintained for some time and then gradually declines in late summer and autumn until it reaches its lowest level in the middle of the winter. The way in which a queen lays eggs in any of the prepared cells she finds within the "brood area" has already been described. She seldom, if ever, wanders outside this area unless she is seriously disturbed—perhaps by someone opening her hive. In the early part of the year, as we have seen, the "brood area" is very restricted in size and the queen is only able to find a small number of cells in which to lay eggs. She therefore does not need very much food from which to obtain the materials necessary for egg production, and as she is only using up the materials already in her body very slowly, she does not frequently solicit the highly nutritious brood-food from the worker bees around her. Gradually, as the mean daily temperature rises, the area of brood comb throughout which the bees maintain the high brood temperature increases, and the bees prepare more and more cells for brood rearing. On finding the prepared cells the queen lays in them; the more cells she finds the more eggs she lays and the hungrier she becomes. The hungrier she gets, the more frequently she solicits food from the worker bees and the greater the amount of food she obtains. Thus colony strength increases slowly at first and then more and more rapidly until it reaches its limit, provided that there are no checks caused by unfavourable weather, bad beekeeping, swarming or other factors. All the time a balance is kept quite naturally between the number of adult worker bees and the size of the "brood area" in which the queen can lay. Later in the season with the advent of cooler nights the size of the "brood area" begins to contract, the queen finds fewer prepared cells in which to lay eggs, and the quantity of brood declines. It is indeed the size of the "brood area" (that is, the size of the area of comb which the bees of the colony maintain at a temperature of 90°F. or higher), and the number of prepared cells available to her, which determine the rate of egg laying by the queen: not, I believe, any deliberate action on the part of her workers.

This is, of course, an idealised and simplified picture of what takes place. Many things can occur which upset the cycle by altering the size of the "brood area" unexpectedly. A heavy honey-flow, and the consequent necessity of using all available cells, at least temporarily, for honey storage, can lead to a marked reduction in the size of the "brood area". Under such circumstances—when the numbers of cells in which to lay her eggs suddenly becomes greatly reduced—the queen, unable to find cells in which to place them, will drop eggs for some time until her rate of output becomes reduced to the required new level. Dropped eggs are greedily seized and eaten by the workers. A queen normally only lays one egg in each cell (Pl. 7a, p. 46), though occasionally when she first begins to lay she will deposit several eggs in each cell in the way that laying workers so often do. After a short time, however, such queens usually begin to deposit their eggs in the normal manner, one per cell.

In the early part of the year the "brood area" of a typical colony is, as we have seen, confined to the centres of a few combs. The queen lays in the cells in this area and, before the resulting bees have completed their development cycle and have emerged, thus releasing these cells for further use, the "brood area" usually becomes extended over a greater part of the comb. A circle of cells immediately outside the already occupied ones comes to lie within the "brood area" and the queen lays in these cells. In this way develop the concentric circles formed of broods of different ages, which are so characteristic of spring and early summer brood nests.

As a laying queen moves over the combs some of the young household bees who happen to be nearest to her at any given time turn to face her and thus form a circle around her (Pl. 19, p. 114). The number of bees forming this circle usually averages about ten. The members of the circle constantly touch their queen with their antennae and lick her body and give her food when she solicits it with outstretched tongue (Pl. 20b, p. 115). This "court" normally consists mainly of young household bees (that is to say bees which have not yet started foraging) and any foragers who happen to meet the queen seem to pay little attention to her. The bees which are paying court to their queen at any moment do not usually follow her around much as she moves, and thus the bees forming the queen's court are constantly changing.

We must now consider the "incubation period" of bees' eggs—the time that elapses between the deposition of fertilised eggs in queen and worker cells, and of unfertilised eggs in drone cells, and the emergence of the adult insects. In each case on the third to fourth day after it was laid the egg hatches, and all three types of cells are closed on or about the ninth day. Whereas, however, queens and workers only spend about five days as larvae, drones usually take six days to complete this stage in their development. Having passed through the pupal stage a queen becomes adult about the fifteenth day after the egg was laid, and usually emerges from her cell on or about the sixteenth day. The worker bee does not emerge until about the twenty-first day and the drone takes longer still: he does not leave his cell until twenty-four days after his egg was laid.

It will be noticed that the lengths of the various pre-adult stages of the Western honeybee, *Apis mellifera*, mentioned above, have been qualified by the word "about" in many cases. This has been done very deliberately and it is incorrect to imagine that, for example, every worker bee has a pre-adult life of twenty-one days, and that every queen spends exactly sixteen days developing in her cell. Such periods as those given must only be regarded as useful approximations, as a rise in temperature of two degrees (particularly during pupal life) can accelerate the rate of development by some hours; and a reduction of temperature can slow development down considerably and prolong larval or pupal life.

It is interesting to note that the average lengths of the pre-adult stages of *Apis indica* are practically identical with those of *Apis mellifera* in spite of the fact that most races of *Apis mellifera* live in temperate climates whereas all those of *Apis indica* live in the tropics. The beautiful temperature regulating mechanisms exhibited by honeybee colonies, involving clustering which maintains the temperature of the "brood area" during cold weather, and dispersal and active ventilation (in extreme cases actually involving the evaporation of water within the hive) which reduces the temperature in this area during hot weather, function so well that "brood area" temperatures are kept remarkably constant in hives in all parts of the world, resulting in almost constant development times.

Normally a queen lives and heads her colony for two or three years, laying as many as 600,000 eggs during this period, but there are many authentic records of queens living as long as four or five years, and sometimes even longer. Eventually, in natural circumstances, when the queen begins to fail, her place is taken by one of her own royal daughters especially reared for the purpose. So, although the individuals succeed one another, the life of the colony is maintained indefinitely.

CHAPTER 7

DIVISION OF LABOUR
AMONGST THE WORKERS OF A COLONY

"Some bees are not (as some conceit) ignorant or negligent in gathering honey, and framing their combs, but yet serviceable in meaner employments (as fetching water, watching and warding) but all are naturally skilful in all their proper labours, and inter-changeably perform them."
SAMUEL PURCHAS (1657). *Theatre of Politicall Flying-Insects*

A T THE PEAK of its strength in early summer a healthy, well-kept colony of honeybees (*Apis mellifera*) consists of anything from about twenty thousand to sixty thousand worker bees (the precise number depending upon the reproductive power of the particular strain of bee, and upon various environmental factors), a few hundred drones and a fertilised queen, together with a number of combs containing brood in all stages of development and stores of honey and pollen.

Some form of division of the many different kinds of work necessary for the colony's well-being must take place amongst the thousands of worker bees, if there is not to be great waste of energy and if the colony as a whole is to function efficiently, as every normal colony undoubtedly does. Ever since man began to study the behaviour of honeybees, and to record his observations, he has been amazed at the wonderfully developed social organisation of the honeybee colony and has assumed, when he has not actually known, that some division of labour occurs. To give but one example, as long as three hundred years B.C. that keeper and close observer of honeybees, pseudo-Aristotle, had recognised the fact that the workers of a colony are divided into a number of different groups at any given time. The members of these groups devote their attentions, most diligently and for the time being, to particular jobs such as building of comb, hive cleaning, water carrying, or the collection of food from flowers. It is obvious, however, that pseudo-Aristotle did not understand how this division of labour is effected. He seems to have supposed that the worker bees are directed to join particular groups of bees that are employed on definite jobs as the need arises. Many of the earlier observers, and even some fairly recent ones, believed that the queen, or as they frequently miscalled her, the king bee, performs this function, directing the labours of the other bees of her colony who are her loyal servants.

Rather fanciful imaginations were also allowed full rein, and even such a generally accurate observer as pseudo-Aristotle was sometimes led astray: for example, he tells us that at dawn one particular bee makes a loud buzzing noise within the hive two or three times in succession in order to arouse the other members of the colony from their slumbers, and send them forth on the duties of the day.

Today we know that a very efficient division of labour takes place amongst the worker bees of every colony. Although this division of labour is brought about automatically (rather than by means of directions given by a superior bee as the ancients supposed) it is very flexible and subject to adaptation to suit the requirements of the colony at any given time. It is not the inflexible, rigid, organisation which some observers have supposed it to be.

Between 1925 and 1930 Dr. G. A. Rösch published the results of some observations which he had made on the division of labour. His were the first really useful data on this subject and were obtained by introducing into a colony a small number of worker bees of known age and marked with harmless coloured paint in such a way that each individual could be recognised at any time. This colony was living in a glass-walled observation hive, and the type of work carried out by the marked bees of known age at different times in their lives was noted. Rösch showed that, broadly speaking, during the active summer months of the year, the adult life of the average worker bee is divided into two main periods. During the first of these, which lasts for just under three weeks, the young worker bee is kept busy carrying out a number of important household duties within the hive, and, during the second period of a further one or two weeks, she is busy in the field foraging. She brings back water, nectar, pollen and, occasionally, a resinous exudation produced by certain plants and known by beekeepers as "propolis". The Western honeybee uses this substance to seal up cracks and crannies in her hive which are too small for a bee to pass through, and some strains of the Caucasian race use it to reduce the size of the hive entrance in winter (Pl. 22, p. 117). (It is interesting to note that *Apis indica*, *A. florea* and *A. dorsata*, as well as some of the races of *A. mellifera* do not make any use of propolis.) Now, it has long been supposed that when a worker becomes adult she possesses a certain, more or less definite amount of expendable energy and that once she has exhausted it she dies: in other words that her expectation of life at any given time normally depends upon the amount of energy that she has already expended or upon the wear and tear connected with her work. It has until recently been supposed that if a bee expended a great deal of energy, her expectation of life will be short quite irrespective of the length of time which she has spent in expending this energy. If, on the other hand, a bee has only expended a small amount of energy it was believed that her expectation of life, barring accidents, would be correspondingly greater.

This theory seemed on the whole to square with the known facts. During the summer months, when bees expend a great deal of energy daily in the field, the length of the adult life of a worker bee is only some four or five weeks; whereas the life of one of her sisters reared during the late autumn may be as long as five or six months, as bees do relatively little work during the cold months of the year. However, two Swiss workers, Dr. Ruth Lotmar and Dr. Anna Maurizio have shown that, although there is obviously a good deal of truth in this wear and tear theory, it is not by any means the whole explanation. They have demonstrated that there are two kinds of worker bees, "summer" bees and "winter" bees which, generally, but by no means exclusively, appear at those times of the year which their names suggest. These two kinds of bees come to differ from one another physiologically and this physiological difference of "winter" bees is partly responsible for their longer lives.

During the first few days of adult life a worker bee normally eats a large quantity of pollen to which she helps herself from the storage cells. This pollen contains protein and enables the young workers' hypopharyngeal glands to develop and to secrete brood-food, and may also result in the slight building up of reserves in certain storage organs, known as the "fat-bodies", within the bees' bodies. If this happens to a bee belonging to a colony rearing a quantity of brood which is in proportion to the size of its population of adult bees (as is usually the case during the summer months) the bee starts to feed the larvae with brood-food and this soon results in exhaustion of any food-reserves stored in her fat-body. Her brood-food glands cease to secrete brood-food actively, and brood-feeding is given up; and the bee soon becomes a forager and wears herself out in the field. Such bees with fat-bodies and brood-food glands which are greatly reduced in size are "summer" bees. On the other hand, when a worker emerges from her cell in autumn she eats pollen in the usual way; but at this time there is often very little brood to be fed. So the food reserves which she accumulates, as a result, do not become quickly exhausted; and her brood-food glands and fat-body only diminish in size slowly, as there is not the great demand upon them which would occur were she feeding brood actively. Such a bee is known as a "winter" bee, and is able to live for months, indeed throughout the winter.

The physiological condition of "winter" bees sometimes occurs outside winter. Anything which causes a substantial and sudden reduction in the number of larvae being reared by a colony, such as loss of the queen, swarming, etc., has an immediate effect upon the young worker bees. Their brood-food glands and fat-bodies will often develop and remain for some time in this condition— i.e. they become "winter" bees—because the requirements of the brood are insufficient to relieve them. In summer (but not in autumn or winter) the situation is relieved by diversion of some of the stored nourishment to the ovaries of the workers; this may lead to the production of laying workers, which, as we have already seen elsewhere (Chapter 5, p. 58), can play an essential part in the continued existence of their colonies under exceptional circumstances.

It is, of course, during the spring and summer months when brood-rearing, nest-building, and so forth are going on most actively, that worker bees are busiest and their lives are shortest. Let us, then, consider the life of an average worker bee at this time of the year.

On emerging from her cell the young worker cleans and grooms her body. As she dries she loses the somewhat bedraggled appearance of the newly-emerged. Within a few hours she is looking her very best, fluffy and well-groomed. Soon after she emerges from her cell, and at fairly frequent intervals during the first three days of adult life, she solicits food from older workers by thrusting her outstretched tongue towards them (Pl. 20a, p. 115). Often her begging is ignored, probably because the bee she happens to ask has no food to give; but she persists until she finds a bee that is able and willing to supply her need. The donor opens her mandibles and moves her still folded tongue slightly forwards and downwards from its resting position tightly folded beneath the head, and regurgitates from her honey-stomach a drop of food which comes to lie on the back of her folded tongue and between her open mandibles. The receiving bee thrusts her tongue into the drop and sucks it up. Both bees use their antennae constantly to feel one another's heads whilst the feeding is going on, in order to keep their respective mouthparts correctly orientated to each other (Pl. 21, p. 116). The food that is given in this way by one worker bee to another consists mainly of nectar or honey regurgitated from the honey-stomach, but may also contain secretions from the brood-food and other glands. This we do not know. It is most probable that, during the transfer of such regurgitated food, the bees concerned perceive any "queen substances" (see Chapter 8, p. 102) with which their partners' heads and their appendages have become contaminated, either directly by contact with the queen herself or indirctly via another bee that has recently been in contact with the queen, and that the transmission (sharing) of these substances amongst the household bees and their perception of them (stimulation by them) is the principal way in which the continued presence of the queen is made manifest amongst the members of her colony. By this I do not mean that those bees that perceive "queen substances" become consciously aware of the presence of their queen in the community, but rather that perception of these "queen substances" by taste and smell prevents the bees becoming restless and their behaviour changing to that which beekeepers describe as "queenless".

According to Rösch the young bee does not help herself to honey from the storage cells in the combs of her nest during the first three days of her adult life, although she frequently does so thereafter. During this period, when not busy grooming herself and soliciting food from other members of her colony, the average bee seems to spend her time either cleaning out brood cells from which adult bees have recently emerged, preparatory to the laying of further eggs in them by the queen, or in remaining more or less motionless on the brood combs and thus helping to incubate the brood.

From the fourth day of adult life onwards the average bee helps herself to honey from storage cells, although she still frequently seeks food from other bees, and often eats a lot of pollen (especially until she is about ten days old) which she takes from the stores within the nest (Pl. 35a, p. 190.) The proteins contained in the pollen that she eats at this period of her life are absolutely essential for the full development of her brood-food (hypopharyngeal) glands. Until these glands begin to secrete, which they normally do about the fifth or sixth day of adult life, the young worker bee is unable to provide the larvae with brood-food. Rösch came to the conclusion that during the third to fifth days of adult life, when she is consuming a lot of pollen the young bee is concerned with the feeding of the older larvae (on a mixture of honey diluted with water, or perhaps nectar, mixed with pollen) but as soon as her brood-food glands become fully functional, on about the fifth or sixth day of adult life, she largely gives up feeding the older larvae and instead begins to feed the younger larvae on brood-food. This she continues to do until she is ten or twelve days old, by which time her brood-food glands have become greatly reduced in size and have ceased to secrete abundantly. In the meantime, however, the wax-producing glands on the underside of her abdomen have gradually become active and, by the time she is about twelve days old, she is ready to begin comb-building and repairing and similar activities which require the production of fresh beeswax. Also at about this time in her life, if weather conditions are suitable, she make the first of a series of so-called "play" or "orientation" flights, during the course of which she learns the position of her home relative to neighbouring objects such as bushes, other hives, etc., and also defecates. The first time that a healthy adult worker bee defecates is during the course of her first "play" flight; and it may well be that the accumulation of indigestible matter in her rectum is the stimulus which causes her to venture from her nest for the first time. No healthy adult worker bee normally excretes within her nest. This first "play" flight is usually taken between noon and three o'clock (G.M.T.) if the weather is warm and sunny.

It is whilst she is between about twelve and fifteen days of age that the average worker honeybee performs the duty of relieving the incoming foragers of the greater part of their loads of nectar, and converting it into honey and storing it away. She also deals with the loads of pollen which returned foragers kick off loosely into storage cells and leave for the household bees to store away properly (Pl. 35b, p. 190). At the same time she does a good deal of house-cleaning, carrying out dead bees and any débris which accumulates on the floor of the hive or nesting hole.

By the time an average young worker bee is about three weeks old she has, during her "play" flights, learned to find her way home again when she ventures into the field. She is ready to take up the duty of foraging for any or all of the things which honeybees collect in the field: water, nectar, pollen, sometimes propolis; and this she proceeds to do until she (usually) dies of old age one or two weeks later. It is usually some of the younger foragers who guard the entrance of their hive or nest against intruders, sometimes in between foraging expeditions (Pl. 29b, p. 162). But most workers omit this duty altogether—otherwise there would tend to be a superfluity of guard bees on duty.

Although many bees perform each of the duties mentioned one

after the other, they do not all do so. Any household duty may be omitted altogether: no observer has been able to demonstrate any fixed sequence of field duties.

The observations made by Rösch on the division of labour amongst the worker honeybees of a colony have both widely and frequently been interpreted as showing that every worker bee in a normal colony carries out a regular succession of duties. It has been suggested that the duty upon which she will be engaged at any given time, is dependent upon a worker's physiological condition at this time; and that, since her physiological condition is closely bound up with her age, what a bee does at any time thus depends upon her age. Rösch (who, incidentally, only observed the behaviour of a relatively small number of bees) never made any such rigid claim. Indeed, he found considerable variation in the age at which the bees he observed became foragers and carried out the various hive duties. His results showed quite plainly that age alone cannot be the determining factor, and subsequent observations by other people have amply confirmed this fact. Thus in his valuable work on the ages of the bees comprising swarms, Morland found that bees of three and four days old are often present. According to those who hold rigid views concerning the division of labour on an age basis, these bees should not have been flying at all!

Observations made by Nelson in 1927 on small, experimentally composed, colonies consisting entirely of very young bees showed that, despite their age, these were able to carry out all the necessary household and field duties. He found that some of these bees started to collect food in the field when only a few days old even

---

Plate 11 (opposite). COMPARISON OF THE SIZES OF THE EASTERN HONEYBEE (*Apis indica*) AND THE WESTERN HONEYBEE (*Apis mellifera*)
Colonies of both of these species of bees naturally build their multi-comb nests in hollow trees and similar protected cavities. As they will live in hives, unlike the Giant honeybee and the Little honeybee, they are the hive-bees of the world.
Above.   Part of a colony of the Eastern honeybee showing queen and workers (x 2.9).
Below.   Part of a colony of the Western honeybee showing queen and workers (x 2.9).

Plate 11

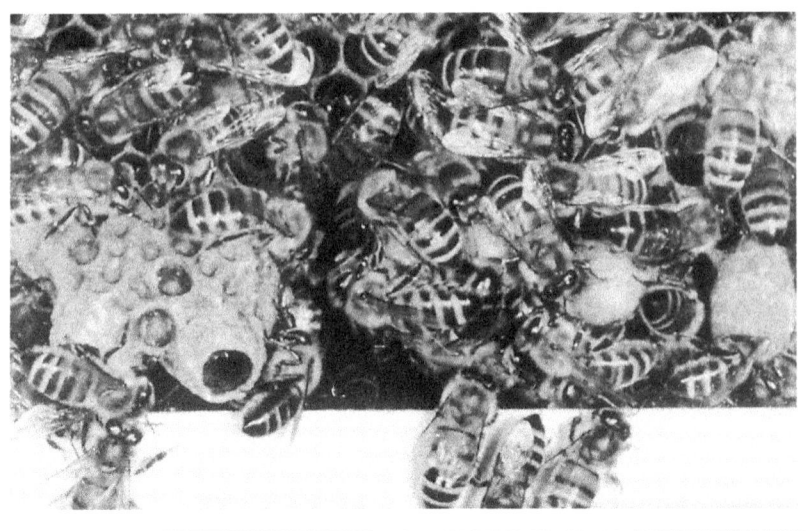

a

b    c

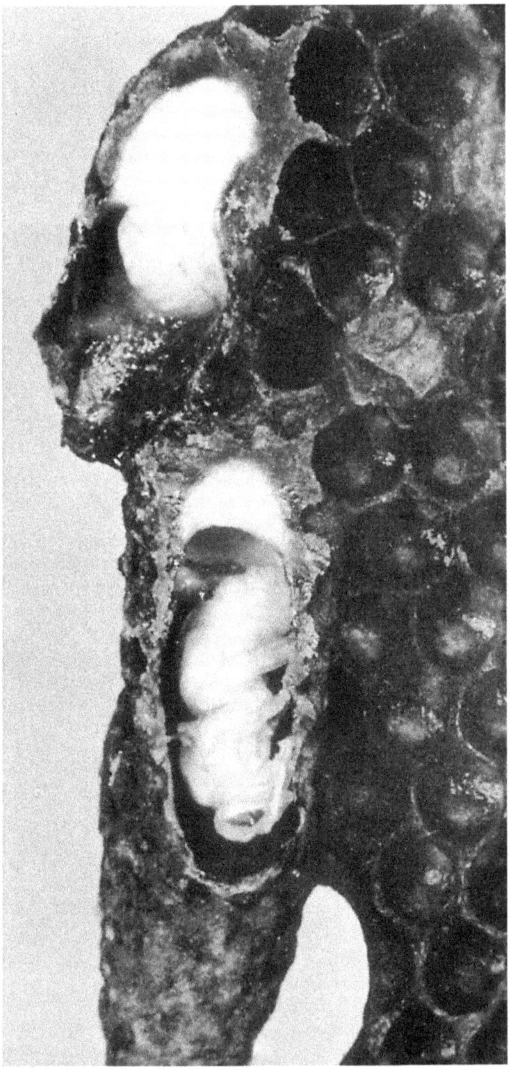

when plenty of sugar syrup or honey was continuously available to the bees within their hives, so that they were certainly not compelled to forage by any lack of honey within the hive.

Betts has described how Himmer observed bees that were only five days old flying from the hive and assisting with hive ventilation and guarding in some of his experiments; he also saw bees that were only eight days old collecting pollen, and bees eleven days old collecting both pollen and nectar. Similar observations were subsequently made by Ribbands who investigated the behaviour of worker bees marked in such a way that they were distinguishable as individuals. In his experiments he found that there was considerable variation in the age at which different bees, emerging from their cells on the same day and living in the same colony (and thus presumably under the same conditions), began to forage. This age varied from nine to thirty-five days!

Himmer also found, in his experiments, that in the absence of suitable young bees, old bees will perform household tasks that are normally carried out by much younger workers. This is well known to beekeepers, who sometimes carry out certain manipulations on their colonies which result in a separation of the older, foraging bees of each colony from the younger, household bees and in which both groups of bees are left with a certain amount of young brood to feed and look after. In one of his experiments Himmer saw bees that were more than forty days old feeding the queen and her brood, cleaning out brood cells—a job usually performed by bees of between one and three days of age—and secreting wax and building combs, as well as collecting food in the field.

It is clear, therefore, that young bees can when necessary

Plate 12 (opposite). SWARM QUEEN CELLS
During the summer months queen cell cups are constructed. Many of these are never used, but a colony when preparing to swarm builds further cups, particularly in groups around the edges of the combs. The queen lays one egg in each cup and the resulting larva is liberally fed with "royal jelly" (brood food) and eventually becomes a queen.

a. Group of queen cell cups along the bottom of a comb (x 1.4).

b. Group consisting of one open and two sealed swarm queen cells on side of a comb (x 2.8).

c. Two cells of group opened to show larva on bed of "royal jelly" and pupa in sealed cell. Note surplus food in base of cell with pupa (x 2.8).

G

W.H.

undertake field duties when only a few days old, and that old bees will secrete wax, build combs, and feed brood when it is necessary for the well being of their colony. No doubt such old bees are unable to carry out these hive duties as efficiently as bees of the age groups which normally tend to perform these functions. It is likely, for example, that their brood-food and other glands will have atrophied to a considerable extent and may not be capable of complete regeneration.

Division of labour amongst any permanent, or semi-permanent, association of worker bees, such as those belonging to a colony, is apparently inherent even when such a group of worker bees possesses neither queen, larvae, nor even combs. A beautiful demonstration of this fact was obtained a few years ago when one of my colleagues released a few hundred worker bees, taken at random from a colony, in a small glasshouse. These soon formed a queenless, broodless, and indeed combless, cluster on the roof. In the course of his work my colleague proceeded to expose a number of small dishes of sugar syrup in the glasshouse; and a small number of bees from the cluster flew down to them, collected some of the syrup, and flew back with it to the other members of their cluster. As each bee settled upon a dish for the first time, she was marked with a spot of paint so that she could be recognised again; and it was found that only a very small proportion of the bees forming the cluster ever went foraging in this way during the ten days or so of the experiment. The bees which did not go foraging stayed in the cluster and depended upon the food brought home by the few foraging bees for their sustenance.

Until about 1950 we owed most of our knowledge about the work carried out in the hive and the division of labour to Rösch, and I have already mentioned some of the many different jobs, particularly household jobs, which worker honeybees perform in a more or less regular sequence during the course of their lives. Now Rösch made his observations by marking a number of bees of the same age and then watching their subsequent behaviour in a glass-walled observation hive. He practised "individual" marking as opposed to "mass" marking but studied what we may perhaps describe as the "average" behaviour of a group of bees of the same age, rather than the behaviour of each of a number of individual bees of known age. In 1952 a research worker who has during the last few years added a great deal to our knowledge of bee behaviour, Dr. M. Lindauer, published the results of some observations which he had made on the behaviour of individual bees, marked directly they emerged from their cells, and kept under regular observation for the rest of their lives. He also used a special type of observation hive in which the bees built their cells with their long axes more or less parallel to the glass walls, thus enabling him to look into the cells from the side. In this way he could see exactly what a worker bee was doing when she thrust her head into a cell.

Lindauer found that the rhythm of life of the household bee (a bee that has not yet become a forager) runs very evenly and smoothly, and that periods of work, or periods when the bee is busy looking for some work to do—patrolling periods—are interspersed with periods of apparent idleness. These periods of work and of presumable rest are in no way related to the time of day or night. Furthermore, should a bee be unable to find work to do in the hive she will quickly give up being a household bee irrespective of her age, and join the foraging force which, of course, can never be too large. Every worker bee spends far more time resting than working; but an apparently idle bee is not by any means a useless bee, as her presence in the brood area helps to maintain the high temperature of between 88°F. and 95°F. that is necessary if brood rearing is to continue successfully. Furthermore, it is possible that the bee's glands are most active whilst she is resting.

Lindauer was able to confirm the general existence of a sequence of hive duties, conditioned by age, as described by Rösch; but the most important of his findings is that this sequence is very fluid, and that each individual bee exhibits a high degree of adaptability to the immediate needs of her colony. Furthermore, contrary to what Rösch and many other workers had supposed to be the case, Lindauer has shown that a bee will carry out several distinct activities, such as the feeding of brood and the building of comb, during one and the same period of her life. Examination of some of these bees has shown that both their brood-food and their wax-producing glands can function simultaneously.

A household bee, a week old, with fully developed brood-food glands, does not go round the brood area looking into cells and only feeding the youngest larvae she can find: she feeds whatever larvae require food, irrespective of their age. She may also clean out any dirty cells she discovers (a job which it was supposed was confined to workers of less than four days old), and she also assists with any comb building and cell repairing which she finds to be necessary. One of Lindauer's marked bees cleaned cells from the day on which she emerged from her own cell until she was twenty-two days old, and she helped to rear brood from the fourth until the twentieth day of her life, whilst she assisted with comb building from the time when she was three days old until she was three weeks old. Lindauer emphasises that this particular bee was not exceptional in her behaviour, and his detailed observations show that the age limits for the occupations just mentioned are very elastic. In other words instead of playing her stereotyped part in some rigid and mechanical system of division of labour, wherein a bee of such and such an age does this job and a bee of another age does that job, the worker honeybee patrols the combs in her hive seeking work, rather than some particular kind of job.

Rösch showed that, as a general rule, a bee gives up household duties in favour of field activities when she is about three weeks old. But should a bee of "household" age be unable to find enough work to do in the hive, she does not idle about at home, but joins the foraging force at a somewhat earlier age than usual. How is this somewhat critical change of occupation accomplished in the life of a worker bee? In the majority of cases it appears to be brought about with the help of successful foragers.

As we shall see later, when we discuss the fascinating subject of the methods of communication between foraging bees (Chapter 16, p. 201), a forager that has found a rich source of nectar or pollen in the field will often dance on the combs when she returns home with her spoils. The dance that such a bee performs is all that is necessary to enable other members of her colony to find out about the kind of food, its floral source, its distance from the hive and the direction in which it lies. It is these dances by successful foragers which help the great majority of the bees that are about to become foragers and enable them to know which

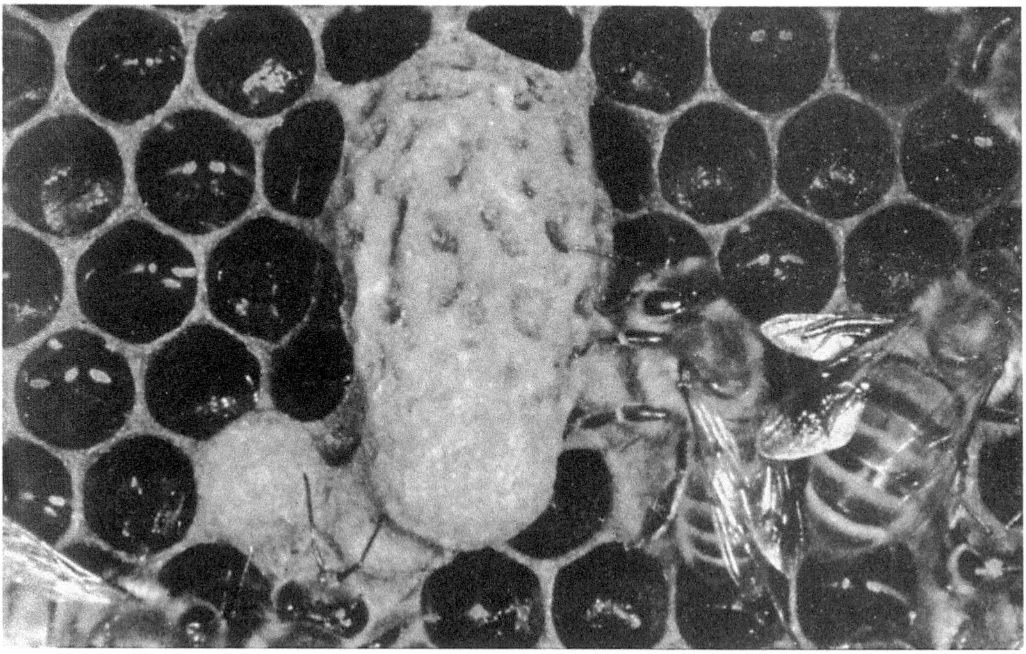

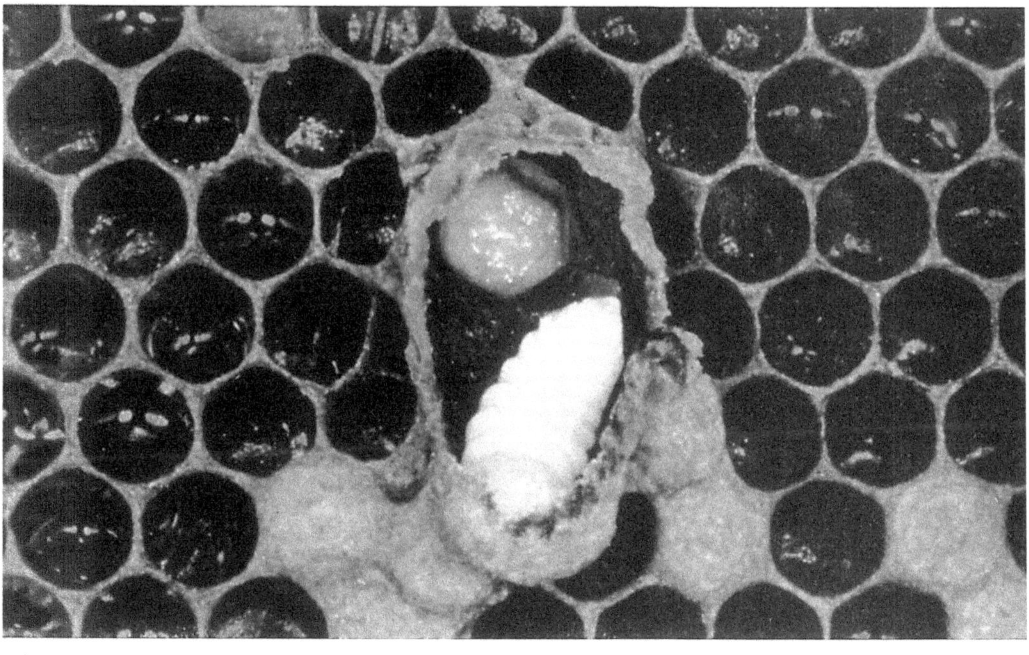

Plate 13.—EMERGENCY QUEEN CELLS

above. Sealed emergency queen cell on face of comb. This might be mistaken for a ' supersedure cell ' until its origin from a worker cell is determined by opening it (x 3·4). *below*. Sealed emergency queen cell opened to show full-grown larva and origin of cell from a worker cell (x 3·4)

Plate 14

kind of flowers to visit in search of food. Thus, of 159 young bees which Lindauer observed, 150 started their foraging lives by following the directions given by older, successful, foragers. They were, in fact, alerted and recruited by the dancing bees. Only 9 of the 159 observed bees appear to have left the hive on their first foraging expeditions without having, immediately previously, followed the dances of successful foragers and learned where to go. Thus, out of these 159 bees only 9 could properly be regarded as scout bees which left the hive, without any preconceived ideas about foraging from any particular kinds of flowers, to seek a source of food quite independently.

However it must not be thought that a household bee, that is about to become a forager, leaves her hive on a foraging expedition immediately she has followed the movements of a dancer. Actually it takes her some hours, or even days, before she is able to follow a dancer's actions sufficiently well to be able to apprehend the information which the various dance movements quickly convey to experienced foragers. At first the recruit follows dance after dance without being able to follow the whole movement correctly as she continually loses touch with the dancer, in the darkness of the hive, when the dancer makes one of the sudden and important turns which are an integral part of the dance. But eventually after a number of attempts which may be spread over many hours or even days, the recruit learns to follow the dancers satisfactorily and thus to obtain the information she requires. When she has done so, she ventures forth to forage and in almost ninety per cent of observed cases returns with food of the same kind and from the same plant as that indicated by the dancer. The recruit may then dance in her turn and so help to recruit and instruct further foragers.

Plate 14 (opposite). SCENT PRODUCTION BY WORKER HONEYBEES (1)
A worker honeybee which has been "lost" on regaining her home exposes her scent gland and fans a current of air over it with her wings, thus disseminating the scent which attracts other members of her colony.
Above.   Number of worker bees exposing their scent glands at hive entrance (x 2.2).
Below.   Worker honeybee in characteristic pose disseminating scent. (Scent gland marked with arrow) (x 3.8).

Lindauer has shown that young foragers know how to dance correctly even if they have never had any opportunity of following a dance themselves. This dance language of the honeybee is, therefore, innate and does not have to be learned, apart from the necessity of the young recruit having to learn to keep in close touch with a dancer throughout her movements, if she is to ascertain the information contained in the dance.

We have seen how even very young bees can become field bees when their services as household bees are not required, and also that (contrary to what has been widely believed until very recently) any household bee is liable to carry out several different types of job within the course of a few minutes. The question therefore arises, does a bee ever carry on both household and foraging duties more or less simultaneously ? This does sometimes happen, and bees that have been foraging have in a few cases been seen to do a small amount of cell cleaning, comb building and even brood feeding, between foraging expeditions. Dr. J. B. Free and I have records of bees, which were seen guarding the entrances of their hives against intruders, going off into the field and foraging, and also disappearing inside the hive for periods of up to an hour (carrying out household duties ?) in between spells of guard duty. It would seem that even those bees of a colony's population who are carrying out household duties, and those who are carrying out field duties, are kept not absolutely distinct from each other. This makes it easier to understand how field bees can, when necessary, quickly readapt themselves to carry out household duties.

Rösch was unable to detect any sequence of duties in the life of a field bee, and more recent work tends to support the view that no such sequence occurs. It is true that, under certain circumstances, a honeybee will for days (and even weeks) continue to collect one particular kind of food from one particular artificial or floral source and visit no others; but recent work carried out in Britain and elsewhere, does indicate that on rare occasions a few bees will forage on one kind of flower at one time of day and on another kind at another time on the same day. A bee will sometimes change on the same or successive days from (for instance) collecting pollen to collecting nectar, and then return to pollen gathering once more.

It is probable that when a bee becomes a forager she does not necessarily, nor indeed even usually, occupy herself with a succession of different duties. The food which she collects is likely to be related to the relative availability of nectar and pollen in the field at that time. When honeybees visit the flowers of wild white clover in search of nectar, a certain amount of pollen is inevitably deposited on their bodies; and the bees gather this pollen together and carry it home along with the nectar which they collect. When nectar is abundant in white clover flowers most bees visiting them bring home, at the end of each foraging expedition, a large load of nectar but only a small load of pollen. But if the nectar in the white clover flowers becomes much less abundant after a day or two (perhaps owing to drought), although many of the same bees continue to visit the flowers, they have to go to many more flowers than before in order to obtain a load of nectar. They will, very probably, collect approximately the same small amount of pollen as hitherto from each flower that they visit, as the pollen supplies probably remain as good as they were before. The result is that each bee, after much hard work, eventually arrives home with a load of nectar and a large load of pollen, instead of the small load she would have collected if she had only visited a smaller number of flowers in order to collect the load of nectar. The type of food which a forager collects at any given time is also determined, to a measurable extent, by the immediate requirements of her colony. Thus it has been found that if a colony has a lot of brood to feed, the foragers collect much more pollen during the course of a few days than they do if the amount of brood is considerably reduced. Similarly a spell of bad weather, during which flight is restricted and pollen stores are diminished, often results in a great increase in the number of bees bringing home pollen loads afterwards.

In other words, when pollen is required bees collect it. But it is not true to say that pollen collection ceases when further supplies of pollen will no longer benefit the colony. It may be reduced but it does not cease; and accumulation of pollen within the hive may lead to a harmful reduction in the number of cells available for brood rearing. It cannot be argued that the bees are compelled to bring home the pollen inadvertently collected when seeking nectar, because it is well known that bees are able to discard pollen in the field and will sometimes deliberately do so. Nevertheless there is evidence that food requirements within the hive determine to some extent the type of food collected by a colony's foragers; though quite how we do not know. Should the colony (for instance) lack an adequate supply of pollen this may perhaps be reflected in the quality of the food that is passing between the adult members of the colony, and may result in the foragers seeking pollen in particular. For a long time bee research workers have supposed that the division of labour within the hive and in the field might perhaps be regulated in some such way but, although scraps of rather circumstantial evidence in favour of such a theory do in fact exist, no really convincing evidence has yet been made available.

If such a theory is to hold, it is first of all necessary to show that an adequate amount of food sharing takes place regularly amongst the adult worker bees of every colony. Now a number of research workers have demonstrated conclusively that food fed to bees within the hive, or collected by them in the field, does become widely distributed amongst both adult and older larval members of the colony in a matter of hours or, at longest, in a day or two. Years ago my colleague P. S. Milne fed a colony with a coloured mixture of pollen and sugar syrup and showed that this mixture was quickly shared amongst both adults and larvae. Another of my colleagues, J. Simpson, and also an American worker, O. W. Park, have shown that when water coloured with a persistent but harmless dye is fed to a colony of bees it can soon be seen in the bodies of workers and in the food of larvae. It was, however, left to another two of my colleagues, C. R. Ribbands and H. L. Nixon, to demonstrate, with the aid of a radioactive tracer technique, how widely and quickly a tiny quantity of food is shared amongst the adult worker bees and older larvae of a small colony. They found that approximately one tablespoonful of sugar syrup, which was collected by only six bees, had within a few hours become widely distributed amongst the adult members of the colony and, indeed, amongst the older larvae as well.

At first sight it looks, therefore, as though food-transmission from bee to bee, which has now been so thoroughly demonstrated, could be a mechanism whereby the bees of a colony are made aware of any surplus or shortage of any particular kind of food, and this might perhaps tend to promote an adequate division of labour amongst the worker bees available. Indeed some years ago, Ribbands sponsored just such a theory based, of necessity, almost entirely upon circumstantial evidence. When, however, we examine each of the cases of food transmission that have been demonstrated, we find that they all have one thing in common: the transmission of food between field bees and from field to younger bees. Even when food is given to bees inside their hive it is probably the field bees who take it from the feeder. Before one can feel fully convinced that food sharing can be the mechanism by means of which the division of labour is brought about, it will be necessary for someone to demonstrate that food is also transmitted from the younger, the household, bees to the field bees. Personally I am doubtful whether this does in fact occur to any extent. It is not enough to show, as has been done, that food is transmitted from field to household bees. It must pass in both directions, more particularly perhaps from the household bees to the field bees, if it is to account for the observed division of labour in the way that has been suggested. At present, therefore, together with a number of other workers, I regard this theory as unproven.

It is nice to suppose, for example, that if protein is in short supply information to this effect is conveyed to the foraging bees by means of food which they receive from the household bees (the bees who would be directly affected by such a shortage) and that the foragers, appreciating the shortage, go off into the field and collect pollen to make good the deficiency. Unfortunately, however, although we know that such protein scarcities in colonies lead to the collection of increased amounts of pollen, we have no reliable evidence as to how this is brought about.

On the whole I am inclined to think that the household and the foraging populations of every colony are more or less distinct from one another, and have little to do with each other, apart from the fact that the foragers pass the bulk of the liquid food which they collect to household bees for any necessary "ripening" and subsequent storage. I feel that the probable link between the household bees of a colony and the foraging bees are the younger foragers—bees which have themselves only recently given up household duties in favour of foraging, that is to say, bees which may be expected to possess recent first-hand experience of the food requirements of their colony. I suppose that every foraging bee will seek nectar (or some similar sweet substance) in the field unless there is some very good reason for it to do otherwise. It can, for example, be imagined how bees that may have felt a partly unsatisfied desire for pollen when they were feeding larvae (or perhaps at some other time in their lives) first of all seek pollen, rather than the customary nectar, in the field when they become foragers. Such pollen-deprived bees might well be quite uninterested in the recruiting dances of nectar-gatherers, but highly interested in those of successful pollen-gatherers, and thus become recruited as pollen-gatherers.

Some support for this theory is to be found in two observations. First, the observation of Lindauer that young foragers belonging to perfectly normal colonies do carry out a certain amount of brood feeding and other household duties between foraging expeditions. So, even though they are foragers, they can be expected to possess a first-hand and up-to-date knowledge of any special food requirements of their colony, and thus to act as a kind of link between the full-time household bees and the older foragers. Secondly: when a colony is short of pollen, potential recruits to the foraging population who are following the movements of dancers (Chapter 16, p. 202) that are carrying pollen will often nibble greedily at the pollen loads on the legs of the dancers, thus suggesting to an observer their own need for pollen.

Probably bees only deliberately (as opposed to inadvertently) collect pollen when they themselves are experiencing or have recently experienced "protein-hunger" (pollen-hunger). It seems probable that "protein-hunger" plays another important rôle in the life of the average worker honeybee. We know that, between the time when her cell is sealed over and the time when she emerges from it as a perfect insect, every worker honeybee uses up almost the whole of the protein reserves which she has previously accumulated from the food given to her during her larval life. It seems reasonable to assume, therefore, that when she emerges from her cell the young worker honeybee is suffering from a protein deficiency—that she feels hungry for protein—and that this urges her to seek out the pollen (protein) stores within the hive and eat large quantities of this substance.

After she has eaten a large quantity of pollen the worker bee's brood-food glands develop so that she becomes capable of feeding young larvae. It may be supposed that the possession of actively secreting brood-food glands causes the individual concerned to seek out larvae requiring such brood-food and to relieve the tension in the storage ducts of her brood-food glands by feeding such larvae. Although this may in general be true, here again we have to remember that in recent work, by Lindauer and others, it has been shown that very young adult workers (too young to possess fully developed brood-food glands) have been seen to feed both young and old larvae. It is also true that young bees with well developed brood-food glands will sometimes take up foraging, if their colony contains a surplus of nurse bees in relation to the number of its larvae which require to be fed, thus redressing the balance, as it would seem that no colony can possess too large a foraging force and accumulate too great a reserve of honey.

Just as pollen collection seems to be tied up in some way with the food requirements of the colony as a whole, which is, of course, made up of the requirements of the individuals of which it is composed, so wax secretion and comb building appear to be linked up with a colony's requirements for comb space in which to store incoming nectar. When a forager returns to her hive with a load of nectar she gives the bulk of it to household bees for processing into honey and subsequent storage. If there is insufficient comb space available for the storage of this nectar, the household bees are compelled to retain it temporarily within their own honey-stomachs. Now it seems that when a bee has to function as a reservoir for nectar for hours on end a fairly high proportion of the contained sugar is assimilated, and that this results in the wax glands secreting wax scales abundantly. Provided that there is enough space in the hive, or nesting cavity, to allow the bees to extend old combs and build further, new, ones they do so. Thus by means of this simple sequence of events sufficient comb space is developed, as and when required, for the storage of "surplus" honey. The whole thing is brought about automatically and most effectively. The greater the quantity of nectar collected, the greater the amount that is surplus to immediate requirements, and the larger the storage capacity required. The larger the storage capacity required in addition to that already available, the longer the household bees are compelled to use their bodies as temporary storage vessels. The longer a bee uses her body for the storage of nectar (or any sugar solution) the more of the contained sugar she assimilates in excess of the quantity necessary as fuel to maintain life, movement, etc. The greater the quantity of surplus sugar that she assimilates, the greater the amount of wax she produces, and thus of the raw material for comb construction.

In the event of the hive having insufficient space for further comb construction the bees will drop the wax scales that are involuntarily produced when they assimilate more than a certain quantity of sugar in a given period of time.

We can conclude that, although the division of labour which exists amongst the worker bees of every colony can no longer be considered to be based strictly upon the ages of the bees concerned, it is nevertheless achieved in an almost equally simple and automatic way.

Three groups of bees can be recognised in every colony, two of which are readily distinguishable from one another, the older, well-established, foragers, and the younger household bees. The third group, which consists mainly of younger bees which combine some work on household duties in the hive with some foraging in the field acts, I believe, as a link between the two groups already mentioned.

A bee on emerging from her cell *always* joins the household group and attempts to carry out any of the household duties which she finds need doing whilst making her tours of inspection. The

---

Plate 15 (opposite). SCENT PRODUCTION BY WORKER HONEYBEES (II)
If a worker honeybee finds a dish containing abundant, concentrated, sugar syrup she may expose her scent gland whilst feeding at it, thus attracting other bees, particularly members of her own colony, to it. Bees belonging to different colonies will feed happily side by side until the supply of food runs short, when they will begin to squabble.

Plate 15.—Scent Production by Worker Honeybees (II)
above. Scent producing bee feeding at a trough. (Scent gland marked with arrow)
(x 3·0)
below. Bees squabbling as supply of food runs out (x 3·0)

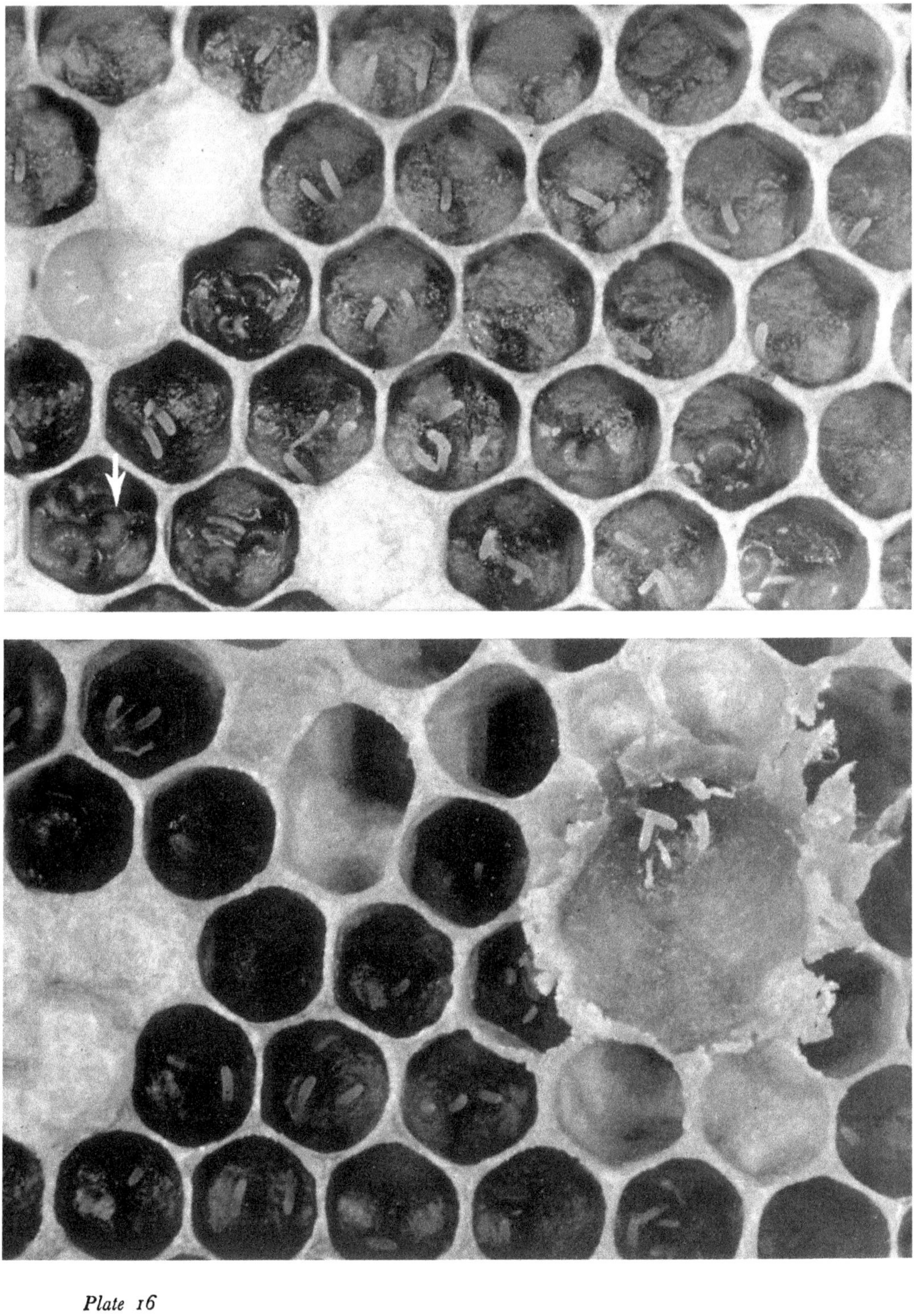

Plate 16

length of time that any individual bee stays with the household group depends, I suggest, upon the amount of work which she finds to do in the hive. The type of household work to which she devotes most of her time will be rather loosely related to her physiological condition and thus, approximately, to her age. As soon as a bee belonging to the household group becomes unable to find sufficient work, of the kind for which she is best suited physiologically, to do within the hive, she probably becomes increasingly interested in the dances of the bees of the well-establish foraging group, or of the "link" group, and sooner or later becomes recruited as a field bee. I suggest that whether she is attracted by pollen dancers or by nectar dancers will depend very largely upon her own condition. If she is "protein starved" she will be attracted by the pollen dancers rather than by the nectar dancers, and so tend to become a pollen-gatherer herself. Otherwise she is likely to exhibit the usual preference for nectar.

When a bee first becomes a forager she quite often continues to do a certain amount of housework and thus to obtain direct information concerning her colony's needs—e.g. if her colony is short of water, or pollen, she feels the lack of a sufficiency of these substances in her own body and this causes her to attempt when foraging to collect the substance that is in short supply, thus helping to make good the deficiency. If a bee is constrained to collect pollen in this way and finds a rich source of this substance, she will dance on returning to her hive and will thus tend, as we have seen, to recruit further pollen-gatherers until the deficiency of pollen within the hive has been made good. Of course it must be realised that even when a colony possesses ample reserves of

---

Plate 16 (opposite). LAYING WORKERS
Laying workers with functional ovaries may develop in a colony deprived of its queen. They usually deposit several unfertilised eggs in a cell but at most only one of the resulting larvae survives to give rise to a small, but fully potent, drone.

Above. Eggs and larvae in worker cells. Note three larvae in cell marked with arrow. Several worker cells sealed with raised "drone" cappings, thus indicating their contents (x 4.2).

Below. Several eggs laid in a queen cell cup. One of these may give rise to a drone or in very rare instances to a queen (x 4.2).

pollen, a certain—sometimes a large—amount will continue to be collected by the foragers, but I believe it is true to say that most of this pollen is collected inadvertently by bees when seeking nectar. Sometimes indeed a nectar-seeking bee who collects pollen inadvertently will refuse to pack it into her pollen-baskets and does not carry it home to her hive.

If neither pollen nor water are in short supply, and if propolis is not required either, a bee will normally collect nectar. In fact nectar collection can, I believe, be regarded as the basic, normal, foraging activity, other substances only being collected deliberately when special need arises.

There is evidence that, as time goes on, a bee becomes more and more confirmed as a forager and presently gives up household duties altogether and concentrates for the rest of her life on foraging.

If a special and urgent need arises for augmentation of the household group of bees, this takes place at the expense of the field force. I believe that bees belonging to the group which I have called the "link" group are, under such circumstances, the first (quite possibly the only bees) to revert to household duties as they are in a position to become aware of the need on account of their part-time hive work. Actually, of course, they find so much to do within the hive that they have little time for further foraging and are also better equipped physiologically for this purpose than older, longer established, foragers.

Thus a remarkable division of labour amongst the worker bees of every colony becomes established, and automatically adjusted whenever conditions change and new needs arise, in such a way that the day-to-day requirements of the colony are effectively met.

Although I do not claim that the existence of the whole of the mechanism which I have described has yet been adequately demonstrated, I believe that this description is likely to be found to be substantially correct; it is surely a very marvellous thing that such an apparently intricate organisation is made to work so efficiently by such very simple means.

CHAPTER 8

RECOGNITION BY A COLONY OF HONEYBEES OF THE PRESENCE OF ITS QUEEN

"They are fond of the person of their queen, whose presence is absolutely necessary to the prosperity and safety of the whole"
JOHN THORLEY (1765). *An Enquiry into the Nature, Order, and Government of Bees*

IF ONE REMOVES the queen from a colony of honeybees the workers soon become aware that she is no longer present in the hive. The state of the colony changes from organised activity to disorganised restlessness. Usually a number of bees will start to run around the entrance of the hive as though seeking their lost queen; and if she is returned on to the alighting-board of the hive they quickly cluster round, touching her with their antennae, offering food and licking her body. Others expose their scent-producing glands and disperse the scent by fanning currents of air over them with their wings. Soon the queen enters the hive escorted by some of the bees that have been running around the entrance, and within a matter of a few minutes the bees around the hive entrance disperse and the colony returns to normal.

Within twelve hours or less, often in three or four hours, the bees of a colony from which the queen has been removed will usually exhibit a much more definite sign of queenlessness. One or more worker brood cells containing young female larvae will be found to have been modified by the bees to form emergency queen cells (Pl. 13, p. 86). The larva in one of these cells is destined, all being well, to become the new queen of the colony.

Let us suppose that we have removed the queen from a colony and placed her in a small wire-gauze cage. An hour or so later we find a number of bees running around in an agitated way on the alighting-board of the hive. We then transfer the queen into another cage and place the cage in which she has been confined on the alighting-board of the hive amongst the excited bees. Their reaction is immediate. They begin to cluster on the cage and many of them examine it thoroughly with their antennae (which, of course, bear organs of taste and smell as well as touch) and lick it, whilst others expose their scent glands and fan. Their immediate behaviour is, indeed, very similar to that which would have occurred had we returned the queen herself, except that as no queen is present they are unable to feed and lick her. At the end of about half-an-hour, however, the bees' interest in the empty cage is clearly dwindling, and after another hour or so the bees are no more interested in it than they are in any other empty cage which has never contained a queen. Obviously the cage in which the queen had been confined had become contaminated with some substance from the queen which greatly excited the bees for a short time, but which gradually became weaker on exposure to the air and to the bees until it was no longer attractive.

So it seems highly probable that the queen leaves some kind of scent in the cage. It is interesting to note that experiments by the author have shown that if one removes a mated, laying, queen from a colony, the bees will apparently be just as pleased to receive a similar mated, laying, queen belonging to another colony as they will be to receive back their own queen, and that any cage in which any queen has recently been imprisoned will prove attractive to excited, queenless, bees. It would therefore appear that all mated, laying, queens share a scent which is very attractive to bees which have just lost a similar queen. One is thus tempted to suppose that bees habitually recognise the presence of their queen by her scent and realise when she is absent because of lack of her scent.

Plate 17 (opposite). DEATH OF DRONES IN AUTUMN
In late summer or autumn when little or no nectar is being collected the drones of every normal colony are driven by the workers on to the combs at the outside of the nest. Here they soon grow weak from cold and lack of food and either crawl out of the hive to die or are dragged out by their sisters.
Above.   Drones (two marked with white spots) being herded together on the outside comb (x 1.9).
Below.   Drone being dragged away from hive entrance (x 2.5).

Plate 17

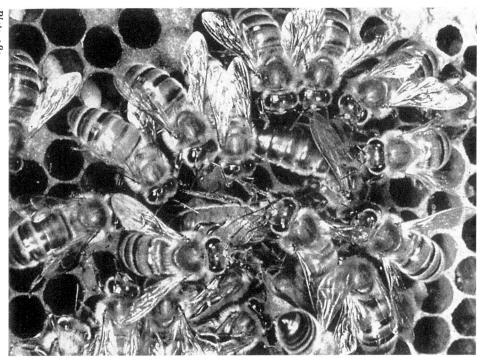

b

Unfortunately this simple theory does not explain everything. For instance, if we place a queen of a small colony alone in a wire-gauze cage amongst the members of her colony, the mesh of the gauze being sufficiently small to prevent the worker bees from entering the cage but large enough to allow them to feed their queen, they seldom appear queenless for 24 hours at least. This observation, which has been made by many different beekeepers at one time or another, fits in with the scent theory reasonably well. Similarly if we take a hive which consists of two brood-chambers, one above the other, both of which contain brood, and separate them from one another with a sheet of queen-excluder zinc (a special zinc sheet perforated with slots of such a size that worker bees can pass through them but the larger queens and drones cannot) so that the queen is confined to one or other of the brood-chambers, we seldom find that the bees in the brood-chamber without the queen show any sign of queenlessness. Here again the scent theory seems to explain matters quite satis-factorily. However, if we now repeat the experiment but this time separate the two brood-chambers with a wire-gauze or perforated zinc screen whose holes are too small to allow a worker bee, let alone a queen, to pass through it we find that sooner or later the bees in the chamber from which the queen is excluded begin to show every sign of queenlessness. And yet there is no doubt that any scent given off by the queen can pass as readily through the wire-gauze screen as it could through the queen-excluder used in the first experiment. It is clear, then, that air-borne scent diffusing throughout the hive atmosphere cannot be the whole answer. Further confirmation of this is obtained if we confine a queen in a double-walled wire-gauze cage, the inner and outer walls being separated by a space of half-an-inch or so, and place the cage amongst the bees of her colony. As in the previous experiment, sooner or later the worker bees grow restless and begin to exhibit signs of queenlessness.

Plate 18 (*opposite*). TWO MATED, LAYING, QUEENS IN THE SAME HIVE (x 2.2). When a failing queen is superseded by a daughter queen the two queens often live side by side in the hive for a time. Both queens are licked and looked after by the workers (*a. left*) and often pass close to each other without any sign of hostility (*b. right*).

Can it be that besides the air-borne scent of the queen it is also necessary that newly laid eggs or young brood be present? This theory breaks down as bees will often begin to show signs of queenlessness within a few hours of being separated from their queen by a perforated screen, even in the presence of plenty of eggs and young brood. Equally the bees in the chamber with the queen will continue to behave quite normally if there are no eggs or young brood present as long as their queen remains with them.

Now, if we reconsider the experiments in which we separated the bees of a colony into two brood chambers, in the first case with a screen through which the bees could pass and in the second by a screen through which they could not pass (obtaining signs of queenlessness in the second case but not in the first) we realise that the most obvious difference between them is that the worker bees could pass to and fro between the two chambers in one case but not in the other. Can it be that in some way or another the bees that pass from the chamber containing the queen into the chamber without her, are able to inform the bees in the latter that the queen is still present in the hive?

I had puzzled over this problem for several years before I finally hit upon what is very probably the solution during the early summer of 1952. At the time I was studying the behaviour of honeybees, both workers and queens, towards strange queens which I introduced into their hives. One day I introduced a virgin queen into a queenright colony housed in a special observation chamber. Almost immediately a number of worker bees seized the luckless virgin and held her firmly whilst others began to form a ball of bees around her (Pl. 23, p. 128). Many of the bees that were "balling" the strange queen in this way made numerous attempts to sting her, but seemed to find great difficulty in so doing and instead, actually stung and killed other members of their own colony who were also helping to form the ball. Presently, although only one strange queen had been introduced, it was noticed that three quite independent balls of angry bees had been formed, one around the virgin and two others around no queen at all. How could the two latter balls be explained? Could it be that the bees were "balling" the virgin queen because they recognised by her body odour that she was a virgin and not a laying queen, and that some of her body odour became transferred to the bees that were "balling" her and that other members of their own colony mistook them for strange, undesirable queens and immediately "balled" and attacked them? Experiments were carried out which demonstrated that this is probably the correct explanation of the bees' behaviour.

Bees from a queenright colony in an observation hive were allowed to "ball" a strange virgin queen and were then re-introduced one at a time into their own colony. In every case the bee was subjected to a detailed examination and some were "balled" and others stung to death, whereas control bees who had not been with strange virgin queens were unmolested when re-introduced to the colony. This suggested to me that the bees nearest to the queen of a colony at any given time, i.e. those forming a "court" round her, probably acquire her odour, or possibly some substance which she either gives to them as food or which they obtain for themselves by licking her body. This substance may become absorbed into the waxes which cover the surfaces of the bodies of worker bees. Whatever this substance may be (and whether it is acquired by the workers deliberately or accidentally, as by contact and absorption) those which acquire it directly from the queen must pass some of it on, either deliber-ately or accidentally, to other bees who in their turn pass it on to others still. In this way, every bee belonging to the colony con-tinues to feel that her queen is present so long as she obtains some of this "queen substance" from time to time, either directly from the queen, or *via* other worker bees.

As the bees of any colony are constantly moving around amongst themselves and feeding each other (Pl. 20, p. 115, and Pl. 21, p. 116) this theory seemed quite plausible: so it was tested by dividing a large colony into three equal parts and transferring bees from that part containing the queen into one of the queenless parts at five minute intervals, using the other queenless part as a control. Three hours after the experiment was started the bees in the control section, to which no bees from the queenright section had been added, were behaving in a queenless manner, and one hour later it was found that three worker cells were definitely being modified as emergency queen cells and that a further two cells were also being prepared. No trace of emergency queen cells could be found in the other two sections. The experiment was later repeated with another colony of bees with similar results.

It appears, therefore, that the bees that were taken from the queenright section and introduced into the queenless section carried with them some substance which they had either obtained directly from the queen, or indirectly from other workers who had obtained it from her: and that the bees in the queenless section on detecting the presence of this substance were satisfied that their queen was still with them and therefore did not build any emergency queen cells. This does not mean, of course, that the bees in this section actually thought about their queen at all, but rather that they detected some of the "queen substance" and were thereby satisfied and inhibited from building emergency queen cells. The bees in the other queenless section to which no bees from the queenright section were added, did not receive regular supplies of "queen substance" and became dissatisfied. Their dissatisfaction was exhibited by those signs which are associated with queenlessness, particularly by the production of emergency queen cells as the inhibitory factor of "queen substance" was no longer operative.

Now it will be remembered that if we take the laying queen of a small colony and cage her alone in a single-walled wire-gauze cage amongst the other members of her colony, the bees feed her through the gauze and seldom exhibit signs of queenlessness for at least 24 hours. If, on the other hand, we cage her, with or without some of her own workers, in a double-walled cage with the wire-gauze walls half-an-inch apart and hang this cage amongst the members of her colony, the bees outside the cage are unable to feed the bees inside or to have any direct or indirect contact with their queen and they soon begin to exhibit signs of queenlessness. This occurs in spite of the fact that with both of these types of cage any scent from the queen can certainly reach the bees outside the cage. It seems clear, therefore, that in the first case the bees received some tangible substance directly from the queen—something which I have called "queen substance" —and in the second case they did not and, therefore, behaved as queenless bees do. Incidentally it does not appear to make any difference to the results of these and similar experiments if we use a mated, laying, queen from a second colony instead of the one from the colony to which the bees belong.

Three questions now arise. First, from what part of the queen does the "queen substance" come? Secondly, what is "queen substance"? And, lastly, exactly how is the "queen substance" transmitted from bee to bee? Let us consider each of these problems in turn. The answer to the first of these questions has been obtained in the following ways.

A small colony of bees, living in a hive which consisted of a single brood-chamber, was taken and divided into two equal parts by means of a bee-tight, vertical, wooden partition right across the hive. The bees in the section which consisted of the front half of the hive could fly, whereas those in the rear section could not. In the centre of this partition there was a hole, covered with a sheet of thin rubber. In the middle of this rubber diaphragm a small, circular, hole was burned and the body of a live queen was pushed through so that her head and thorax protruded on one side of the partition and her abdomen on the other side. Thus her head and thorax were exposed to the bees on one side of the partition, and her abdomen to the bees on the other side. The bees of both groups paid a great deal of attention to whichever part of their queen they could reach. Those in the section into which her head was protruding fed her and frequently licked her head and thorax and examined them with their antennae. The bees in the other half of the hive, into which the queen's abdomen was protruding, removed her excrement and any eggs which she produced and also examined her abdomen with their antennae and licked it. Neither group of bees built any emergency queen cells.

It is clear, therefore, that the "queen substance" can be obtained by the worker bees either from the head and thorax, or from the abdomen, of the queen. It is not necessary for the bees either to feed the queen or to remove her excrement in order to obtain sufficient of the "queen substance". That it is, however, necessary for the bees to touch the queen is shown by the fact that after the bees which could only reach the queen's abdomen had shown no signs of queenlessness for three days, they started to build emergency queen cells a few hours after the abdomen of their queen had been covered with a wire-gauze cage to prevent them from touching it.

Ultimately it was found that the "queen substance" is obtainable by worker bees from all parts of the body surface of their queen. In one experiment it was found that access to three legs and about four square millimetres of the thorax of their queen was sufficient to prevent a small group of bees from building emergency queen cells. Anything which interrupts, or seriously diminishes, the availability of "queen substance" amongst the worker bees of a colony results in removal or reduction of the inhibitory effect of this substance and the building of emergency queen cells.

Thus evidence suggests that when a queen becomes old or ill, and failing, her production of "queen substance" decreases and her workers promptly either build emergency queen cells or tolerate and maintain supersedure queen cells. A "supersedure" queen cell differs from an "emergency" queen cell, and has a similar origin to a "swarm" queen cell. It is constructed as a queen cell from the beginning, not modified from another kind of cell, and the queen actually lays an egg in it from which is produced a queen who will ultimately supersede her as head of the colony.

It is also probable that a breakdown in the collection and distribution of "queen substance" occurs in a colony which is going to swarm and results in the bees of such a colony tolerating eggs and larvae in queen cells preparatory to swarming (see p. 162).

In 1953 I was able to demonstrate the quantitative nature of the "queen substance" in an experiment which has been successfully repeated a number of times. An apparatus was devised by mean of which a mated, laying, queen in a wire-gauze cage could be automatically placed amongst the bees of a section of her colony for a predetermined length of time and then transferred (still in her cage and without any worker bees) to another, separate, section of her colony, where she again remained for a predetermined period of time before being transferred back into the first section and thence to the second and so on. A colony of bees was moved in its hive to a new site a few yards away from the old ones. The field bees were allowed to drift back to a trap hive, containing a little brood, placed on the original site. The queen was caught and placed in the cage of the apparatus. The young adult bees were then divided into three groups, each of which possessed combs containing both food and larvae of all ages. One group was placed in a hive and kept as a control; the other two groups were placed in the special experimental hives into which the queen in her cage could be automatically placed for pre-determined periods of time. The mechanism was adjusted so that the queen remained in one of these hives for six minutes in every hour, and in the other hive for fifty-two minutes each hour, the remaining two minutes being the time when the queen was in process of being transferred from one hive into the other. In one such experiment about 4,000 bees were used in each of the experimental hives as well as in the control hive. The bees in the experimental group which had the queen for fifty-two minutes per hour built three emergency queen cells in forty-eight hours, those which had her for six minutes built ten, and those forming the control group which did not have their queen with them at all built twenty-seven. It seems clear, therefore, that a given number of bees must receive a certain minimum quantity of "queen substance" in any given period of time if they are to remain inhibited from building emergency queen cells.

In 1954 two Dutch workers, Professor A. P. de Groot and Dr. S. Voogd found that development of the ovaries of groups of queenless worker honeybees in cages could be inhibited by allowing them access to material they extracted from queens' bodies in ethanol. In other words they showed that development of laying workers (see Chapter 5, p. 57) in a colony is normally prevented, like queen rearing, by material produced by the queen —i.e. by "queen substance". Later, my colleague, Miss D. Gibbons and I, similarly found that ethanol extracts of whole queen honeybees inhibited queen rearing by queenless worker bees in cages when given to them either on the bodies of dead worker honeybees or in their drinking water. Soon afterwards, Dr. J. Simpson and I demonstrated that the principal sources of this "queen substance" are the queen's mandibular glands—two large glands in her head just below her eyes. We presumed that this secretion becomes dispersed over the surface of the queen's body, probably when she is grooming, and so available to the worker bees of her colony. This, then, appeared to be the answer to the first question—from what part of the queen does the "queen substance" come. However, about twelve years later, another Dutchman, Dr. H. H. W. Velthuis, produced evidence suggesting that, although the mandibular glands are the principal source of "queen substance", other glands in the queen's abdomen also probably produce the same, or other, inhibitory materials.

Our discovery that "queen substance" is produced in a queen's mandibular glands greatly helped my colleagues, Dr. R. K. Callow and Dr. N. C. Johnson, and me to identify at least one of the major components of "queen substance". It is a fatty acid called 9-oxodec-*trans*-2-enoic acid, which, as shown by biological and chemical tests, becomes spread over the queen's body. Synthetic "queen substance", 9-oxodecenoic acid—became available in quantity in the winter of 1960-1961, thanks to the generosity of Glaxo Laboratories Ltd., after Drs. Callow and Johnston had made the first synthesis of it. We then found that, contrary to expectation, 9-oxodecenoic acid alone does not cause complete inhibition of queen rearing and that at least one other substance, the scent of another fatty acid, 9-hydroxydec-*trans*-2-enoic acid, which is also produced in the queen's mandibular glands, is involved. It is, therefore, more correct to refer to "queen substances".

Although complete inhibition of queen rearing by queenless worker honeybees has been obtained many times in cages by giving the bees extracts of queens or the secretion from their mandibular glands, and a high level of inhibition has been obtained in cages with synthetic 9-oxodecenoic acid, and with mixtures of 9-oxodecenoic acid and 9-hydroxydecenoic acids, no one has yet succeeded in achieving complete and prolonged inhibition of queen rearing in large colonies of bees with anything except living queens.

Finally we come to the third question—how are the "queen substances" transmitted from bee to bee? We know that something of the sort happens because several experimenters have shown that queen rearing by worker bees and development of their ovaries can be inhibited without the workers themselves having any direct contact with a queen, provided these workers have ready access to other workers that have recently been with a queen. So far we have assumed that, some of the "queen substances" acquired by a bee, either directly from the queen (perhaps by licking her body) or from another bee who has obtained it either directly or indirectly from the queen, are passed to other bees in the food that is exchanged, for it is well known that when bees meet one another they examine each other with their antennae and often one of them then offers regurgitated food to the other. So it does seem to be possible that "queen substances" are transferred from one bee to another in such food, and quickly shared among the workers of the colony. As no satisfactory method of showing experimentally whether "queen substances" are passed from bee to bee in this way could be devised, recourse was had to direct observation, and data were thereby obtained which strongly suggested that this is the way in which this is achieved. If a laying queen is watched in an observation hive as she moves over the brood combs, it will be seen that some of the household bees that are nearest to her on the comb at any given time turn and face her and form a group or "court" around her. Many of these bees examine her body with their antennae and some also lick it with their tongues (Pl. 19, p. 114), whilst others feed her, probably with brood-food from their hypopharyngeal glands (Pl. 20b, p. 115). Observations were, therefore, made on the actions of twenty bees in turn, each of which was seen to lick the body of her queen, and also of the actions of another twenty bees which examined the queen with their antennae but did not lick her body. Each bee was watched continuously for five minutes after she had left her queen having either examined and licked or only examined her. It was found that every one of the bees seen to lick the queen offered food to at least one other member of her colony within five minutes of leaving the queen, often within a few seconds of doing so, and the majority fed several bees in succession. On the other hand, not one of the bees seen merely to examine the queen with her antennae was observed offering food to any other bee within five minutes of leaving the queen. It seemed highly probable, therefore, that some of the bees in the "court" round the queen were obtaining "queen substances" by licking her and subsequently sharing some of them with other members of their colony. A bee that has been in contact with a queen will often travel a considerable distance in the hive during the next few minutes and contact, and sometimes feed, a number of bees. This suggestion of mine that inhibitory "queen substance" is obtained by worker bees licking their queens and then shared by them with other bees in regurgitated food, did not, of course, exclude the possibility that other, perhaps more important, methods of transmission of inhibitory "queen substances" between workers may exist. Reconsideration of the results of some of my own observations and consideration of data subsequently collected by Dr. Voogd and by Dr. Velthuis, leads me to conclude that this is the case.

I have found that my early observations of the proportions of the bees in "courts" that examined and licked their queens, in contrast to merely examining them with their antennae, and that fed other bees shortly after leaving a queen, were abnormally high. It is also clear from the work of Dr. Voogd that the effectiveness of inhibitory "queen substances" becomes greatly diminished when they are mixed with concentrated sugar solutions such as occur in worker bees' honey-stomachs. It is probable, therefore, that a worker bee that examines a queen with her antennae, or one that examines and licks a queen, not only has her antennal sense organs stimulated by "queen substances" but also, as Dr. Voogd suggested, herself becomes contaminated with traces of these inhibitory substances which, during feeding and similar encounters, stimulate the antennal sense organs of other bees inhibiting them from queen rearing, and also perhaps contaminate their bodies so that they in turn can stimulate other bees. It is likely that "queen substances" pass from queen to worker, and then between workers, by contamination during contact and that these substances act through a recipient bee's brain rather than directly on her metabolism. Time lapse photography has shown that the movements and contacts that are continually going on in a hive are fully adequate to ensure thorough mixing of the household bees of a large colony in a short space of time and thus, one presumes, sufficient sharing of "queen substances" to inhibit queen rearing. The field bees appear to pay very little attention to their queen and it seems doubtful whether they are interested in "queen substances" or require them.

There is no doubt that although both virgin and mated, laying, queens are attractive to worker bees, the latter are able to distinguish between the two very quickly. The workers of a colony headed by a mated, laying, queen will usually form a ball around any virgin queen who enters their hive and attempt to kill her (Pl. 23 p. 128), whereas they will seldom immediately attack another mated, laying, queen. It appears likely, therefore, that there is some difference in the quality of the materials produced by virgin and mated, laying, queens which enables bees to distinguish between them. It seems certain that they are not distinguished by their behaviour, as even when their behaviour on being introduced to a strange colony appears to be identical (as for example when both are recovering from the effects of an anaesthetic) a clear distinction is made between them by the bees. The formation of balls of angry bees around worker bees contaminated with virgin queen substances which was described earlier, will also be remembered in this connection. In no other connection has the author ever seen worker bees "balled" by angry groups of stinging bees, although he has seen many robber bees stung to death by the guard bees of the colony whose hive they were attempting to plunder (Pl. 28b, p. 155.) The "balling" of strange queen bees is a quite distinct phenomenon from the mauling and killing of intruding worker bees.

Some substance on the body of an intruding virgin queen, in addition to her strange body odour, is sufficient to release the stinging behaviour of the bees of a colony headed by a mated, laying queen, even when the intruding queen makes no attempt to escape when challenged by them and is entirely submissive. The work of Drs. Yadava and Smith in Canada indicates that much, if not all, of this substance comes from the mandibular glands of the unfortunate virgin queen; being released by her in response to the approaches of the workers challenging her. By contrast, the strange body odour of an intruding worker honeybee together with any mandibular gland secretion she releases is (as we shall see when we consider the question of colony defence, Chapter 12, p. 143) insufficient to arouse the stinging responses of the guard bees and she is most unlikely to be stung, provided that she remains submissive and does not attempt to escape or appear to the guards to be attempting to do so.

It has long been assumed that a mated, laying queen attracts the workers that form a "court" around her by means of an odorous substance she produces, perhaps in her mandibular glands. However, time lapse photography seems to show that normal "court" formation round a queen—as distinct from aggressive "balling"—probably involves little, if any, orientated movement of workers towards the queen from a distance of more than a few millimetres. It seems that some of the household bees that are moving around the brood-nest meet the queen more or less accidentally. Many of those that do so appear either to ignore her or even to be repelled by her and quickly move away, but some seem to be strongly stimulated by her and these join her "court", examine her body and often lick it too. When the queen moves many of the bees in her "court" lose contact with her but a new "court" is soon formed by other bees elsewhere in the brood-nest when the queen stops.

When some of the worker bees of a colony that has been without a queen for an hour or so find a queen, they not only tend to stay with her but also to disperse scent from their Nassanoff glands (see Pl. 14, p. 87), so encouraging other bees to move towards them and so towards the queen. This seems likely to be the explanation of observations that have been made of the worker bees of disturbed colonies—e.g. when swarming—moving towards their queens from some distance away. So far as we know worker bees in a queen's "court" do not attract other bees either with Nassanoff scent or in any other way.

CHAPTER 9

CLUSTERING AND THE MAINTENANCE OF COLONY COHESION

"A bee, like a man cannot live alone, if she be alone she dies: . . .
Bees conserve community unto their last: . . . They express if not
great reverence, yet I am sure great love to their Commander: . . ."
SAMUEL PURCHAS (1657). *A Theatre of Politicall Flying-Insects*

A NORMAL COLONY of honeybees during the summer months may consist of 30,000-40,000 worker bees, a few hundred drones and a single queen. It is obvious that if each worker bee were to act quite independently of all the other members of her colony the result could only be chaotic. A division of labour exists amongst the worker bees of every colony, and we have already discussed a number of facts concerning it and various theories that have been put forward to explain them. We have also considered the probable steps taken by bees along the evolutionary road from a purely solitary existence to the highly developed social organisations exhibited by the honeybees today. We have not yet, however, discussed the probable ways in which cohesion is maintained amongst the members of a honeybee colony.

Although a honeybee can live for some time on her own, provided that conditions are suitable—for instance in a properly equipped cage in an incubator in the laboratory—no individual bee will willingly do so. Honeybees are very dependent upon other members of their kind and a queen honeybee is quite incapable of founding a new colony on her own, unlike a queen bumblebee, wasp or ant. She has become so highly specialised for egg production that she has lost the ability to collect food, build a nest and feed larvae.

The great attraction that any group of worker bees has for any bee that has got "lost" has been studied in France by Lecomte, and later in England by Free and myself.

If a hundred or more worker bees are taken and scattered about in an empty box in the dark they will form a cluster together within a few hours. But if, as Lecomte was the first to show, less than about fifty bees are placed in the box the chances are that they will remain scattered or only form small, temporary groups, each of not more than three or four individuals.

Now, if two cages made of wire-gauze are taken, one being empty and the other containing a couple of hundred well-fed bees, and placed six inches apart in a darkened arena, and then another two hundred bees are scattered in the arena, the latter bees will probably form two clusters, one on either cage, simply because the cages form convenient supports. If only one suitable support is provided a single cluster is always formed. However, let us suppose that the bees have formed two clusters, one on each cage. Presently as the bees outside the cages become hungry they get restless, and those that have clustered on the empty cage begin to migrate from it to the cage containing the well-fed bees and join those bees that are already clustering there. Thus, in time, a single cluster is formed on the occupied cage, and the hungry bees of which this cluster is composed solicit food from the well-fed bees inside the cage. Even if the bees inside the cage are themselves fairly hungry the bees outside still tend to cluster on their cage, and will still do so if the cage has double walls through which they are unable to make any physical contact with the bees inside. But if the bees outside the cage are provided with plenty of food, they are just as likely to form a cluster and remain on the empty cage as they are upon the one containing the other bees. It is clear that the reason that the bees go to the occupied cage, in preference to the empty one, is because they are hungry and wish to seek food from the bees inside it. Here is one of the chief factors which serve to hold the bees of a colony together.

As their social organisation has evolved, the honeybees have developed the practice not only of the adult workers feeding the larvae but also of adults feeding other adults. Thus workers give nectar, syrup, or honey, to other workers and to drones, and brood-food to queens as well as to larvae and, as we have seen (Chapter 8, p. 108), "queen substances" are probably transmitted from bee to bee during food sharing. We can, therefore, conclude that interchange of food is one of the factors responsible for the maintenance of colony cohesion.

A second important factor is the bees' need for warmth. The honeybee is a cold-blooded animal and her body temperature when she is resting approaches that of the air immediately around her. However, whilst she is undertaking vigorous muscular activity, such as flying, her temperature rises considerably. As soon as a bee's body temperature falls below 46°F. she becomes chilled and almost incapable of movement. On sunny days in winter, bees often leave their hives on "cleansing" flights* when the air temperature is only a few degrees above freezing point. So long as they continue to fly they are all right, but should they alight on the ground, or remain on the alighting-board of their hive for any considerable time they become chilled and incapable of rejoining their companions, and they ultimately die.

Between the middle of November and February honeybee colonies in Britain contain very little, if any, brood, and the temperature of the general atmosphere of the hive is almost the same as that of the outside air. For weeks it is so low that any isolated bee would very quickly become chilled. However, the bees do not remain widely distributed in the interior of the hive but, instead, form a compact cluster on a small number of combs and in the spaces between them. The change from summer conditions, when temperatures are relatively high and the bees are fairly uniformly distributed throughout the hive, to winter conditions, when the bees are clustered together, occurs when the temperature of the general hive atmosphere falls to 56°F. or so. The nearly spherical cluster formed of broodless bees provides considerable thermal protection, and the temperature of the air in the inner parts seldom drops below about 68°F. and usually varies between approximately 68°F. and 86°F. throughout the

---

*Flights during which they excrete, a function that honeybees do not perform within the hive except when suffering from dysentery.

period. These temperatures, which are of course much higher than outside temperatures, were at one time supposed to be maintained by the bees deliberately undertaking vigorous muscular exercise, such as by fanning their wings or wagging their abdomens from side to side, whenever the temperature within the cluster dropped below the region of 70°F.; and it was suggested that a regular interchange of positions took place between the colder bees forming the outer walls of the cluster and their warmer companions on the inside. Actually there is no reliable evidence in support of this view and plenty against it. It seems that these high temperatures are produced automatically as a by-product of the metabolism of the bees that are forming the cluster and of any brood present.

Cluster temperature is maintained by contraction and expansion of the cluster itself in accordance with changes in the temperature of the environment. When the bees first start to cluster together on the advent of colder weather in the autumn the cluster is only loosely formed, but as the cold becomes more intense the bees pack themselves more and more closely together, thus maintaining the cluster temperature by lowering the rate of loss of heat by reducing the surface area of the cluster itself.

Some time after Christmas, usually early in February, brood rearing begins again and a much higher temperature of between 90° - 95°F. is usually observable in the centre of the cluster where the brood is situated. However, even in such brood-areas temperatures as low as 88°F. have been recorded in winter in Britain. Unfortunately the factors responsible for initiation of brood rearing early in the year are not yet fully understood.

This ability of colonies of the Western honeybee (*A. mellifera*) to maintain, by means of clustering, sufficiently high temperatures for survival even in the cold of subarctic winters, coupled with their capability of withstanding high temperatures, has made perennial existence of the colony possible, and has also played a

---

Plate 19 (opposite). A LAYING QUEEN SURROUNDED BY A GROUP OF ATTENDANT HOUSEHOLD BEES

The queen in this picture is being fed by the bee marked x, licked by the bee marked y, and examined by several other bees (x 2.8).

Plate 20.—THE FEEDING OF ONE BEE BY ANOTHER (I)
This behaviour plays an important part in colony organisation and cohesion.
above. Worker bee on comb face feeding another worker (x 2·8).   *below.* Queen
accepting food from one of her workers (x 2·8)

most important part in their successful colonisation of so large a part of the world.

Since warmth is so important to the honeybee it is not surprising that it was found that when workers were scattered inside an arena in the dark at a temperature of 65°F. they were attracted to, and clustered on, a tin whose temperature was maintained at a slightly higher level than that of the air in the arena. It is clear, therefore, that the individual's need for warmth is another important factor in the maintenance of colony cohesion.

Probably the most important single factor helping to keep the members of a colony of bees together is the transmission of "queen substance" between the workers, at all events between the household bees (Chapter 8, p. 102). Their need for "queen substance" is readily shown by the following experiment in which two colonies of bees, each with its own colony odour (p. 141), are caused to join together, a thing that they would not normally do. Two hives containing combs of food are arranged so that their entrances lead into a small, darkened, arena. A group of worker bees from a colony (queenless or queenright) is placed in one hive without a queen and a group of bees from another colony together with their queen (mated or unmated) is placed in the other hive; within a few hours, the queenless bees will quit their hive and move across the arena to join the queenright group in the other hive. There is evidence that the bees of the queenless group learn of the presence of the queen in the other hive because some of those who venture into the arena obtain "queen substance" from bees of the queenright group who happen to be there also. The queenless bees then give some of the "queen substance" to other members of their own colony, and gradually the whole lot are attracted across the arena to the hive of the queenright bees.

The workers of some kinds of ants and termites have been found to obtain a "queen substance" from their queens and are inhibited, like honeybees, from rearing further queens as long as the supply is sufficient. It is probable, therefore, that the need for "queen substance" is a very powerful force indeed towards the maintenance of colony cohesion in each of these groups of social insects. There is no evidence that the workers of the wasps and bumblebees obtain any particular substance from their queens,

but, among the wasps at least, there is evidence that they obtain
something from the larvae which they imbibe greedily; perhaps
this may act in a way similar to "queen substance" in keeping
the members of a colony together.

We must now consider the senses which come into play when
an individual worker bee is attracted towards other worker bees
in a darkened arena.

The worker bee's ability to realise that a place or an object is
warmer than she is herself probably plays a part in this connection.
A semi-chilled worker will tend to be attracted towards a cluster
of bees whose bodies are warmer than her own.

Yet another most important factor is the bee's sense of smell.
The odour of other bees, particularly that produced by their scent
glands, is extremely attractive to isolated bees. Free and the
author have been able to demonstrate this by means of a specially
constructed olfactometer in which individual worker bees were
caused to run up a tunnel down which two streams of air were
moving at equal speed, one of these streams being drawn over
the carefully excised scent glands (Pl. 14, p. 87) of other worker
bees. At that part of the tunnel where the two streams of air
entered side by side, with the minimum of turbulence and mixing,
the bee had to choose between one or other stream and follow
it for a short distance. A significantly greater number chose the
stream that was being drawn over the exposed scent glands. It
made no difference to the result whether the scent glands over
which it was drawn were obtained from bees belonging to the same
colony as the experimental bee or to another. These results indicate that bee scent is strongly attractive to other bees from whatever source it is derived.

Obviously another sense which may be expected to play a
part in helping a bee that is "lost" in a darkened arena to find

---

Plate 21 (opposite). THE FEEDING OF ONE BEE BY ANOTHER (II)
The bee who offers food to another regurgitates a drop of liquid food so that
it comes to lie on the back of her still folded tongue and between her open
mandibles. The bee receiving the food does so with extended tongue.

Above.   The worker bee on the right is feeding the worker on the left (x 3.5).

Below.   The worker bee on the left is offering the food (x 3.5).

Plate 21

Plate 22

other bees, especially other groups of bees, is the sense of hearing, provided that a bee possesses it.

Many of the earlier writers supposed that honeybees are able to appreciate the same sort of sounds as man does, and indeed that they have a language of sounds. Thus, for example, one reads that when some of the members of a colony realise that their queen has been taken from them they commence to moan, or wail, and by these sounds to communicate their distress not only to the beekeeper but also to the other members of their colony. In recent years, however, this highly anthropomorphic view has tended to be discarded, especially as a number of workers have failed in attempts to train honeybees to associate certain sounds with the presence of food, or to frighten bees with sounds produced at the hive entrance with violins, whistles, etc. Furthermore the great authority on honeybee anatomy, Dr. R. E. Snodgrass, in his book on this subject (*The Anatomy and Physiology of the Honeybee, 1925*) stated that he could not find any organs in honeybees which he could definitely regard as hearing organs, functioning in a similar way to those of, say, grasshoppers and crickets.

As we have already seen (p. 62), queen honeybees sometimes behave in a way that produces shrill, piping sounds to which it has been reported that other queens nearby reply, although they cannot see the piping queen. Dr. Simpson has shown that the piping sounds are produced by operation of the flight motor without the wings being spread and that they are transmitted through the comb on which the active queen crouches to nearby queen and worker bees; the workers responding by "freezing". It has also been shown that worker bees in an observation

---

Plate 22 (opposite). THE USE OF PROPOLIS
Propolis consists of a mixture of different resins which bees collect from plants, a good deal being collected from the sticky buds of poplar and horse-chestnut trees. Only races of the Western honeybee collect propolis and use it for sealing up cracks and crevices in the hive, or sometimes to reduce the size of the hive entrance in winter.

Above.   Worker bees bringing home propolis. The bee on the left still has some in her pollen baskets (x 3.5).

Below.   Bees blocking up the holes in the perforated zinc mouse-guard across the entrance of their hive (x 3.5).

hive will respond in the same way to the vibrations that are produced when one rubs a moist finger to and fro on the glass wall of the hive, or touches the hive with a suitable, vibrating tuning-fork. Even a person singing will sometimes cause workers to "freeze". Results of experiments like these show that workers also respond—though not necessarily to the sounds which we hear, which may be quite inaudible to them. Again, although we know that some insects, such as crickets, are able to appreciate vibrations which reach them through the air, there does not appear to be any really convincing data to show that even queen honeybees are able to appreciate airborne vibrations, as distinct from those which reach them through some other medium such as the comb upon which they happen to be standing.

The results of some of Lecomte's experiments suggest that individual worker honeybees in a darkened arena will respond to some extent to the vibrations produced in the walls of an opaque cage by the movements of bees contained in it; but he was not successful in showing that such vibrations alone were sufficiently attractive to cause individual bees to settle near their source. However, in other experiments, Free and the author have been able to show that bees are indeed attracted by vibrations produced by other bees, and even by the vibrations of the walls of an old cocoa tin containing an electrical vibrator working off the alternating current of 50 cycles per second. Thus it is probable that the vibrations produced by the bees of a cluster are either "felt" or "heard" by individual bees nearby, and help the latter to find the cluster.

Lastly there is, of course, the question of sight. All the experiments discussed so far were carried out in darkness, or with individual bees out of sight of other bees, so as to exclude any possibility of the isolated individuals being attracted by the sight of other bees moving.

In an experiment in the light it was found that bees are not attracted to groups of dead bees, even if they have only just been killed. This result agrees with one in which a number of individual bees in turn had to choose between currents of air. The experimental bees showed no preference for a current drawn over the bodies of a number of freshly killed bees. On the other hand they showed a marked preference for currents drawn over the excised scent glands of bees from any source. It seems, therefore, that most of the scent produced by worker bees comes from their scent glands. In the case of even a freshly killed bee the scent gland is almost unexposed and what little scent is available appears to be insufficient to attract other bees, nor does sight of her do so.

In other experiments in the light, glass-walled cages containing living bees, and others containing no bees, were suspended about six inches apart from one another about a glass-covered arena in which a number of bees had been scattered. The scattered bees showed a tendency to cluster beneath the occupied cage as they might have been expected to do. That worker bees under more natural conditions are attracted by the sight of others has been conclusively demonstrated in the field by Kalmus. He trained bees to visit a certain place to collect food from a number of glass dishes placed on pieces of white paper. He then found that on approaching the feeding place these bees always tended to select from a number of otherwise similar, clean dishes (used in order to eliminate any trace of residual bee scent), one in which he had imprisoned two or more live decoy bees beneath a piece of glass. A few years ago one of my colleagues and I found that social wasps, who apparently produce little or no scent (and who unlike honeybees, appear to have no means of communicating information about the position of foraging sites to one another) are also attracted to sources of food by the sight of other wasps flying around or feeding at these sources.

We can conclude, therefore, that it is probable that the most important factors in the maintenance of colony cohesion are, food interchange, the need for warmth and the need for "queen substance"; whilst the smell of a cluster of bees, the warmth of it, the vibrations produced by it and the sight of it, all help to lead individual "lost" bees to join it.

CHAPTER 10

THE WORLD
OF THE WORKER HONEYBEE

"Bees have all the five senses, tho' there do not appear all those outward organs of scenting, which some other animals have. . . ."
JOHN GEDDE (1721). *The English Apiary or the Compleat Bee-Master*

BEFORE we can hope to understand very much about the behaviour of the honeybee and the way she reacts to stimuli in the world around her, we must consider very briefly the special senses with which she can appreciate things in her environment. It is, of course, foolish to suppose that a bee sees the world as we see it, and indeed, the results of research work are continually showing how very different the bee's world is from our own.

For example, although our ancestors probably possessed the ability to do so, how many of us are able to realise our proximity to a supply of pure water, however closely we approach to it, if we are unable to see, touch, hear or smell it? But, as Hertz has demonstrated, worker honeybees can do so.

Hertz found by experiment that bees were attracted to dishes whose contents gave off water vapour even though they could neither see the contents nor reach them with their tongues. On the other hand the bees did not alight on dishes which contained oil, glycerine or alcohol, none of which give off water vapour. Thus it became clear that the bees were able to detect the water contained in some of the dishes although they could not see it.

It must not be supposed, however, that worker bees can "smell" pure water in the same way as they can smell a perfume. Indeed all the data indicate that they are unable to do so;

nobody has yet succeeded in training bees to the scent of pure water, although it has been found possible to train them to associate a very wide range of perfumes with food. No, almost certainly honeybees have sense organs which are directly stimulated by the water molecules themselves, and which enable them to detect slight local increases in the water vapour content of the atmosphere immediately around a hidden water supply. These sense organs are probably located on the bees' antennae.

In nature it is, of course, very unlikely that a honeybee will ever come across a supply of distilled water. The sources of water which bees tend to frequent usually contain many dissolved substances and have pronounced, and often (to man) unpleasant odours. Such sources of water are, as beekeepers know, to be found in small, stagnant, ponds which contain dead leaves and other decaying organic matter; in leaf-choked gutters (Pl. 34, p. 183); in the small puddles which sometimes collect on cow-dung pats, and in the effluent from manure heaps. Clearly if the bee can smell the odours associated with such sources of water, and has learned to associate them with water, they will aid her in her search for it. Some years ago, in early spring, the author trained a large number of bees to collect water from dishes placed on a table in the open. Water from various sources, as well as solutions of a number of individual salts and mixtures of salts, and distilled water, were made available in the dishes and any preference which the bees showed for one or other of them was noted. It was found that they preferred water containing small amounts of certain salts, such as common salt and ammonium chloride, to distilled water, but that they greatly preferred cow-dung-water (water collected from puddles on cow-dung pats), sewage effluent, etc. to any of the salt solutions offered. It seemed as though the bees were more interested in the malodorous substances in such sources of water rather than in any salts which they contained. In order to check this point some cow-dung-water was filtered and then distilled. The salts and other non-volatile substances which remained behind after all the water and volatile substances had been boiled off, were redissolved, as far as possible, in a bulk of distilled water equal to the distillate containing the volatile substances. The bees' liking for the distillate, which smelt like the original cow-dung-water, was compared with their liking for the redissolved salts, and for some of the original cow-dung-water, as well as for distilled water. It was found that the bees exhibited no significant preference for either the original cow-dung-water or the malodorous distillate, but that they greatly preferred either of these sources of water to distilled water or to the solution of residual salts.

Some of the distillate was then taken and much of its odour was removed with charcoal. Further tests then showed that the bees' liking for it had become greatly diminished. Thus it was concluded that worker honeybees seeking water visit the puddles on cow-dung pats (and probably also many other sources of water which smell unpleasant to man) not on account of any dissolved salts which they may contain, but rather because of what appears to be, from the bees' point of view, their most attractive smell. Free has shown that bees often disperse scent on finding unscented water.

Again, experiment has shown that worker honeybees possess the ability to note the passage of time remarkable accurately. Probably the first recorded experiments on this subject were those of Forel who, in his book *The Senses of Insects*, published in 1908, tells how one day he noticed a solitary honeybee stealing jam from his table, and how the next day several bees, presumably recruited by the bee that he had seen the day before, were visiting the jam dish. Now Forel only had jam on his table twice each day, at breakfast and again at tea, and he noticed that after a day or two the bees only arrived at breakfast and tea time, and never at lunch or supper time when no food was available on the table for them. It looked as though the bees did not just happen to see the table laid for a meal and come down to it as they had learned that they might find food on it, but rather had learned that food was only available to them at certain definite times each day, and therefore, only came at these times. In order to make certain that this was so, Forel conducted a simple experiment. He arranged that no jam was put on the table at any meal. The bees arrived at breakfast and tea time as usual and, failing to find the jam dish, started to hunt for jam in the cups and other pieces of crockery. The next day also no jam was put out, but still a few bees arrived at breakfast time and rather more at tea time, but, in addition, a few bees came at lunch time, a meal at which no food had ever been available to the bees.

Since Forel's early experiments a great deal of work has been done in attempts to elucidate the mechanism of this well-developed time-sense of the honeybee. Nevertheless, although many interesting data have been collected we do not yet know exactly how the bee tells the time, or rather marks its passage. She certainly does not do so by recognition of any regularly recurring factor in her external environment, such as the position of the sun, although such environmental factors do play an important part in the system. For instance, Beling was one of the first to show that when a colony of honeybees was kept continuously in a room in which the light intensity, humidity and temperature were all maintained at constant levels throughout the twenty-four hours the bees could be trained to visit a dish at any time of day, even in the middle of the night. But, under these artificial conditions they "told the time" less accurately than they did when living in the open air. As we shall see later this latter observation helps us to understand a little more about the mechanism concerned.

Because none of the external factors of the environment which she studied appeared to be of primary importance in enabling honeybees to note the passage of time, Beling and many other workers have explored the possibility that this ability may depend upon some internal metabolic rhythm within the bee's own body. Much evidence in favour of such a theory has been collected. Some of the first was obtained by Kalmus who showed that when a bee, that has been trained to visit a dish of sugar syrup which is made available to her for a definite and limited time each day, is subjected to prolonged cold (about 5°C. for nineteen hours) during an interval between feeding times, thus temporarily reducing her rate of metabolism by slowing down her internal processes, she will arrive slightly late at the feeding place. Similarly, Grabensberger has shown that when a bee trained in this way is fed with a drug (euquinine) which decreases her metabolism she arrives late at the feeding place. On the other hand, when she is fed with another drug (iodothyroglobulin) which increases her rate of metabolism, she arrives early at the feeding place. In both these cases no food was made available to the bees on the test day, nor on subsequent days. In each experiment the bees made errors of several hours on the test day, but on the following day they arrived at the feeding place more nearly at the correct time, and their accuracy was even greater on the third day. Clearly in some way their "clocks" had been corrected. (It must be noted that no food had been exposed at the feeding place since the bees had been given the drug so that they had no opportunity to relearn the correct feeding time). These results seemed rather unexpected. One might have thought that once the bees had been caused, by an induced change in the rate of their metabolism, to arrive too early (or late) at the feeding place they would subsequently have continued to make an error of the same magnitude. Later on, Lindauer and some of his colleagues in similar experiments found that when trained bees were narcotised for between two and five hours with carbon dioxide they subsequently visited the dish of syrup to which they had previously been trained mainly during two periods, one corresponding to the visiting time to which the bees had been trained, the other being later, the extent to which it was late being roughly equal to the duration of narcosis.

The results of these and other experiments strongly suggest that the time-memory of the honeybee is based on at least two independent mechanisms which can be separated experimentally, but are normally synchronised with the natural day. The first of these is an endogenous time-measuring system, which appears to be independent of external factors and rigidly bound to a period of rather less than twenty-four hours (i.e. about 23.4 hours), whereas the second system is controlled by external factors, such as the position of the sun. A spectacular demonstration of the working of the honeybee's time-memory system was obtained when a colony of bees was transported overnight from Paris to New York. The next day the bees arrived at the dish of syrup at their original, Paris, feeding time, then, in the course of the next few days, became adapted to the new suntime.

It has been found that bees can be trained to visit selected feeding places during two or even three different periods on the same day. However, if the intervals between feeding times are made too short (less than about two or three hours) the bees tend to fly around the feeding places all day long. This accurate time sense of the worker honeybee appears to be innate and, together with her ability to learn to seek food in a certain place and to associate particular floral colours, patterns and perfumes with food, is of great importance to her, as some kinds of flowers produce plenty of rich nectar, or an abundance of pollen, only at certain more or less definite times each day. The bees soon learn to conserve their energy by only visiting such flowers at those times when they know, from past experience, that the flowers are likely to contain worthwhile quantities of food. A bee may even visit one kind of flower at one time of day and another kind in a different place at another time on the same day. Bees will often remain in their hives in between visiting times.

Von Frisch has shown that a foraging honeybee also measures "distance", or, more correctly perhaps, "time spent in flying", between the hive and a source of food. He has suggested that this sense of "dead reckoning" is also dependent upon the bee's metabolism and that she "measures the distance" by appreciating the amount of energy—the quantity of fuel—she has had to expend on her outward journey from the hive to a particular source of food. She is able, on returning to her hive from a successful foraging expedition, to give this information to her sisters (by means of the bees' dance language, Chapter 16, p. 201) when recruiting them to exploit the rich source of food that she has found. Von Frisch has noted that when a bee flies out from her hive against a strong head wind she subsequently indicates, in any recruiting dance which she may make, that the "distance" from hive to food is greater than she does after visiting the same source of food on a calm day. Similarly, if she flies out with a following wind behind her, her dance appears to suggest that this "distance" is less than it would have been had there been no wind. Whether or not von Frisch's "metabolic" theory is the correct explanation of the observed facts, the ability of a honeybee to measure "distance" is of great importance both in helping her to find her way about in the field, and in providing would-be foragers with information about sources of food she finds (Chapter 16, p. 204).

A high proportion of a bee's life is spent within the almost complete darkness of her hive; it would therefore seem that touch is important in enabling her to find her way around. What do we know about the honeybee's sense of touch ? In order to study this sense use has been made of simple mazes through which individual bees have had to find their way in order to reach a source of food, the whole maze being kept in total darkness.

It does not take most worker bees very long, even in complete darkness, to learn the correct route through a simple maze, with perhaps two blind alleys, to a dish of syrup entirely by touch. The maze has, of course, to be thoroughly cleaned between runs in order to remove any scent trail which the bee may have left behind her. Individual bees differ markedly in their learning ability and some never seem to be able to learn the route really accurately. The bees have to learn the way back from the food to the entrance of the maze quite separately from learning the way from the entrance to the food. The same thing has been found to be true of ants. It is clear, therefore, that, despite their un-doubted learning abilities, these insects are unable to make any use of the experience gained on the outward journey through the maze to help them on their return journey.

Something of the sensitivity of a bee's organs of touch, pre-sumably those of her feet, is shown by the fact that even in total darkness she can learn to distinguish between strips of glossy cardboard, tin, glass and other materials. If strips of cardboard are placed at those junctions in a darkened maze at which the bee should turn left, and strips of metal at those at which she should turn right, a bee will learn to run the maze accurately much more quickly than she would do without these aids. Bees will very quickly learn to follow a cardboard track laid across the floor of a darkened, glass, arena from the hive entrance at one side to a dish of food at the other. Every time a bee runs off the cardboard strip she makes circling movements until she finds her way back on to it again.

This delicate sense of touch is situated in her antennae and feet, and probably in other parts of her body as well. This sense together with an acute sense of smell (the organs of which are located on her antennae), are all that seem to be necessary to enable a bee to find her way about quickly in the darkness of her hive.

How can we write so definitely of the way in which a honeybee behaves in the dark ? Fortunately a study of colour vision of the honeybee shows that she is unable to perceive a deep red colour. If we carry out our maze and similar experiments in light of this colour we are enabled to watch the bees, admittedly with a certain amount of difficulty, whilst they for their part can see nothing at all. If it were not for this very fortunate physiological fact we should not know nearly as much about the behaviour of the honeybee as we do.

If the worker honeybee is blind to red, can she see other colours ? Even before the first World War Lubbock and Forel had carried out experiments which showed quite conclusively that bees can distinguish between differently coloured papers. Unfortunately, however, neither of these great naturalists designed their experiments in such a way as to exclude the possibility that the bees were distinguishing between the different papers which they used, not by differences in colour as such, but rather by the relative brightness of the different colours to the eye of the bee.

It was von Frisch who first carried out experiments which demonstrated that worker honeybees can distinguish between colours as such. To do this he exposed a number of pieces of cardboard of different colours on a table in his apiary and amongst them he placed a number of pieces of cardboard of different shades of grey. He found that he could train his bees to seek food from, and distinguish between, orange, yellow, green, violet and purple cards, and to distinguish any of these coloured cards from grey cards of any shade. But the bees would persist in confusing red with dark grey or even black cards, showing that they are colour-blind to red. Thus we find that worker honeybees have the ability to distinguish between coloured cards by their colour rather than by their relative brightness. Had the bees, in von Frisch's experiments, been distinguishing between the coloured cards by differences in their brightness they would, of course, have got them mixed up with some of the grey cards of equal brightness.

As a matter of fact, von Frisch found that although the worker

bees were able to distinguish between the various colours mentioned, they often made mistakes. For example, bees will confuse
blue cards with violet and purple ones, and yellow cards with
orange and green ones. In fact if bees are trained to seek food
on orange cards only, and are then offered green and yellow
cards as well, they actually prefer the green and yellow cards to
the orange ones to which they have been trained. It therefore
seems that to the eye of the honeybee yellow, orange and green
all appear as the same colour sensation, but yellow is the most
saturated colour of the three and appears to her to be more
striking and attractive than the others. Similarly blue, purple
and violet are sometimes confused with one another and, of the
three, blue and purple are more attractive to the bee than violet.

Even after these experiments of von Frisch had been completed
the possibility still remained that the bees might perhaps be
distinguishing between the coloured cards, not on account of
differences in the colours as we see them, but rather because the
cards of different colours might be reflecting more or less ultra-
violet light, which is, of course, invisible to man. It was improbable perhaps, but not impossible. Ultimately this last
difficulty was overcome by Kühn in 1927 who carried out similar
experiments using spectral light itself. He confirmed all von
Frisch's conclusions and also discovered that bees are able to
distinguish blue-green from blue and yellow, and to see ultra-
violet as a true colour to which they can be trained. From these
experiments we find that, whereas man can distinguish between
over fifty different colours, the honeybee can almost certainly
only recognise four distinct qualities of colour, those which we
call yellow, blue-green, blue and ultra-violet.

What does all this mean to the bee? Let us try to see; but in
so doing we must remember that whereas we cannot see ultra-
violet light the honeybee can, and that although we can see red
light the bee cannot. Thus the brilliantly scarlet flowers of the

Plate 23 (opposite). THE "BALLING" OF A VIRGIN QUEEN
If a strange virgin queen enters the hive of a colony which possesses a mated,
laying, queen, the workers seize her, form a ball around her and attempt to
sting her to death.

Plate 23.—THE 'BALLING' OF A VIRGIN QUEEN (x 2·8)
above. A few bees have seized the Queen
below. The 'ball' of bees is forming around the strange Virgin Queen

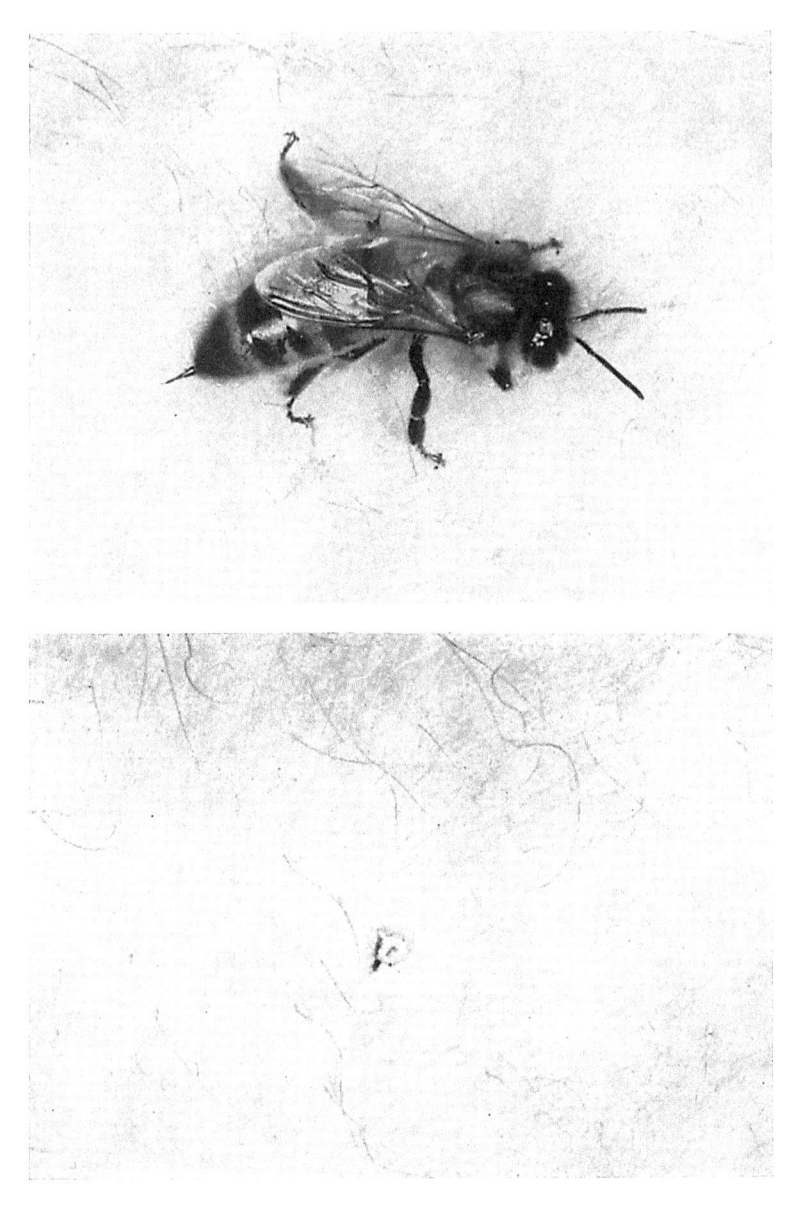

Plate 24.—STINGING

Whereas bees can withdraw their stings from the bodies of other bees they cannot do so from human flesh. A bee who loses her sting in this way dies prematurely

above. Bee attempting to withdraw sting from back of man's hand (x 3·0)
below. The sting left in the hand after the bee has torn herself away (x 3·0)

field poppy, from which bees often collect pollen, reflect a lot of ultra-violet light and probably appear bluish to bees. Many other red flowers have a touch of purple about them and almost certainly appear to bees as some shade of blue.

Now, some white objects absorb ultra-violet light whilst others reflect it. This means, of course, that two objects which appear equally white to us may appear as of two quite distinct colours to bees—the first can readily be distinguished by them probably as a shade of blue-green, whereas the other kind of white object probably appears as a shade of white.

Almost certainly (as in the case of the human) in order to give the bee the sensation of white "colour", all the colours which her eye can distinguish have to be present in the same proportions as they are present in sunlight. If this is true, it follows that, since the bee can see the colour ultra-violet, light of this colour must be present together with the four different qualities of colour which the bee can see—that is, roughly speaking, light of every wavelength which we can see except red—if the bee is to receive the colour sensation produced by whiteness to man. If the ultra-violet light is absent the object still appears white to man, whereas in the case of the bee the wave-lengths of light which are present combine together to give her the sensation produced by the complementary colour,which is probably blue-green. Therefore, since most white flowers absorb ultra-violet light, such flowers probably appear to be blue-green to bees.

In nature the colour, shape and perfume of a flower in which she has found food are soon memorised by a bee, and this knowledge greatly aids her in her subsequent foraging. Of all the characteristics of flowers the two which are of the greatest importance to honeybees are colour and perfume. We shall see something of the parts these play in attracting bees to flowers when we come to consider foraging behaviour.

The precise shape of a flower does not seem to play an important part in attracting bees, probably because they lack the ability to observe detail when flying along.

The large, compound eye of the worker honeybee consists of over six thousand separate elements, each of which registers a minute spot of light corresponding in brightness to the average of all the light values within its field of view, which is about one degree. These spots of light together form a mosaic, in much the same sort of way as a picture in a newspaper is made up of a mosaic of black and white spots. It has often been stated that the honeybee is very short-sighted, but this is incorrect. It is true to say, however, that the honeybee cannot see fine detail, simply because only a relatively small number of her eye elements can view an object at any one time.

As a foraging bee passes an object as she flies along, different elements in her compound eyes, which are firmly fixed in her head, view the object, and this must result in a flickering impression. The greater and the more frequent the contrasts made by the object with its background, the greater the stimulatory effect on the eye of the bee. This probably explains why it is that bees that have learned to associate particular flowers with food, on returning to seek more food, find and settle upon the flowers more quickly when they are moving in a light breeze than they do when the flowers are stationary on a calm day.

The honeybee's sense of smell was very thoroughly investigated, many years ago, by von Frisch. He placed on a table in his apiary a number of small cardboard boxes, each of which had a small hole in one side to serve as an entrance for the bees. One of the boxes contained a small dish of sugar syrup and also, on a little ledge just over the entrance, a few drops of scent such as oil of orange. All the other boxes were empty.

The position of this "training" box containing the food and scent, in relation to the other boxes on the table, was changed from time to time to prevent the bees from learning to seek food from a box in one particular position on the table. When a number of bees had learned to visit the training box in search of food all the boxes were removed and replaced with clean ones, none of which contained food and only one of which contained a trace of the scent. The bees soon returned and flew around the boxes, paying particularly attention to their entrances, but did not attempt to enter any of them until they found the one containing the scent, whereupon they entered this box in search of food. Clearly the bees had learned to associate the scent with food. The results of von Frisch's many experiments on the worker honeybee's sense of smell indicate her ability to learn to associate any one of a large number (44) of perfumes with food and to distinguish it from the others.

Von Frisch also came to the conclusion that some substances of different chemical composition but similar odours to man also appear to bees to have similar odours, and that some substances that are odourless to man are likewise odourless to bees. He also concluded that the lowest concentrations of various scents which can be appreciated by man are also about the lowest that can be perceived by bees. This was rather a surprising conclusion when one considers the relatively low sensitivity of the human sense of smell compared with that of many animals. Modern man does not as a rule rely very largely upon his sense of smell to find his food as do many animals, including the honeybee. It is not altogether surprising, therefore, that when the thresholds of perception of scents by the worker honeybee were recently reinvestigated by Ribbands, who used pure, scented, chemicals, rather than scents extracted from flowers as von Frisch had done, it was found that these thresholds were much lower for bees than for man.

In bees the olfactory organs are located on the antennae and, as the latter are extremely mobile, can be brought into very close contact with objects which their owner is examining so that even traces of a scent are discernible. This is important not only when a bee is examining a flower but also in enabling her to discern perfumes of flowers on the bodies of foragers that have returned home after finding a rich source of food. Furthermore, it enables a bee, that is examining another bee with her antennae, to distinguish a member of her own colony from a stranger from another colony by a slight difference in their body odours (Chapter 12, p. 147).

When we turn our attention to the worker honeybee's sense of taste we find that she can taste sweet substances, such as sugars, with sense organs on her feet as well as with others on her antennae and mouthparts. Many, indeed most, substances which appear sweet to us are apparently tasteless to the honeybee. Indeed it is almost true to say that whereas most sugars appear sweet to man, only those kinds which commonly occur in the natural liquid food of the bee, nectar and honeydew, appear sweet to her.

K

W.H.

Such artificial sweetening substances as saccharine seem to be tasteless to the honeybee at concentrations at which they are very sweet to man, and repellent at higher concentrations.

It has been shown by Wykes that the bee's relative preference for certain sugars varies with different concentrations of the sugar solutions. The preference exhibited by the bees in all her experiments for solutions containing equal proportions of sucrose, glucose and fructose, is unexpectedly high when considered in relation to the relative preferences shown by bees for other solutions of sugars. This result probably has some biological significance since nectar, the principal natural source of carbohydrate for the bee, usually consists of a mixture of sucrose, glucose and fructose.

So we find that the world of the honeybee must be very different from our own. It is a world of silence, as the worker bee is almost certainly completely deaf, in the sense of being unable to appreciate airborne sounds. It is a world in which the movement, colour and perfume of a flower are more important than its shape; in which cornfields glowing with scarlet poppies appear as a rather dull shade of blue, and in which many white flowers are blue-green, and some white painted hives also appear as this colour whilst others are white; a world in which there are no reds but plenty of blues and greens, a world of four colours, yellow, blue-green, blue and ultra-violet. It is a world in which the sun is useful as a compass (Chapter 16, p. 199) but almost unheeded as a clock, where time is measured in terms of some internal metabolic rhythm, and distance possibly by the length of time during which certain body-hairs are bent; where scent is all-important in finding food, in recognising home, and in distinguishing between friends and foes. It is a world dominated by one love, that of the queen—because of the substances she produces (Chapter 8, p. 109); and in which, as shown by von Frisch, the sight of a tiny scrap of blue sky is enough to place the position of the sun extremely accurately (Chapter 16, p. 208). Most surprising of all, it is a world in which companions exchange information by dances.

CHAPTER II

BEE STINGS AND BEE VENOM

"When you are stung, or any in the company, yea though a bee have stricken but your clothes, specially in hot weather, you were best be packing as fast as you can."
CHARLES BUTLER (1623). *The Feminine Monarchie. 2nd ed.*

MANY FEMALE INSECTS possess an organ known as an ovipositor. Some, such as the crickets, use their ovipositors for thrusting their eggs into the ground, whilst others, for example some of the sawflies, use them to make slits or holes in leaves into which they insert their eggs. The ovipositor of the female honeybee, both queen and worker, has lost this function, and has become modified as a weapon for defence, the sting. Simpson has shown that a queen honeybee will continue to lay eggs in a normal manner even after her ovipositor has been cut off.

Queen honeybees can sting human beings, but only very rarely do so; indeed they are only inclined to attack other queens. Virgin queens, even those belonging to the same colony, will nearly always start to fight as soon as they meet one another and will continue until one of them is killed. A virgin queen, if introduced to a laying queen belonging to another colony, will attempt to sting the latter to death; but, in the natural course of events, if a virgin queen wanders into the hive of a strange colony she is balled (Pl. 23, p. 128) and killed by the worker bees before she has time to attack the colony's own queen. On the other hand, in many cases of supersedure a virgin queen is produced by a colony whilst the old, laying, queen, her mother, is still present and the old one often fails to attack her. Whether this is because she has no inclination to do so, or whether she is somehow prevented from so doing by the worker bees, we do not yet know. Frequently a virgin queen reared under the supersedure impulse will mate and begin to lay whilst her mother is still present; sometimes they are found almost side by side on the comb (Pl. 18, p. 99) and showing no trace of animosity towards each other.

When a queen honeybee attempts to sting another queen, or a worker bee attempts to sting either a queen or another worker bee, she often seems to have great difficulty in driving her sting home. It is probable that a bee can only penetrate the body of another bee with her sting through one of the relatively soft parts of the body, such as the intersegmental membranes which act as hinges between the rigid, harder, parts. It is not easy to find out exactly whereabouts in a bee's body the sting is introduced because, contrary to widespread belief, the victor removes her sting from the body of her victim at the same high speed as she thrusts it in. In fact the motions of thrusting in the sting and removing it follow so quickly upon one another as to be almost continuous, and take place more quickly than the eye can follow. It is only rarely that one finds a bee's sting in the body of another bee.

Some people seem to doubt whether bee venom is effective against other bees, but there is no doubt that it is. Bees that have been stung by other bees crawl away obviously mortally injured and usually die within a minute or two.

As a rule worker bees will only attack human beings and other animals in defence of their colonies and homes, except, of course, if they have been picked up, or otherwise molested, and are attempting to escape. When a human being is stung the worker bee is in the great majority of cases unable to withdraw her sting from her victim's flesh and in her struggles to do so (Pl. 24a, p. 129) soon ruptures the delicate tissues by which it is attached to her body and breaks away leaving her sting and its associated structures behind (Pl. 24b, p. 129). A bee that has been damaged in this way may survive for a few days, but not for long.

The sting is normally hidden inside a cavity, the sting-chamber, at the tip of the bee's abdomen, from which it can be quickly protruded when necessary. A detailed and very clear account of the structure and mode of functioning of the sting and its mechanism has been given by Snodgrass. The following, very condensed, account is based on his work.

The sting itself consists of a straight, tapering, tube, the base of which is swollen to form a bulb, and is formed of three pieces which together surround a central canal. One of these lies dorsally and forms a kind of sheath to the other two parts (ventral lancets) which lie beneath it and slide to and fro longitudinally on track-like ridges of the sheath. These two ventral lancets are provided with barbs at their extremities. The central, or poison, canal is formed from a groove, along the underside of the sheath, which is closed below by the two ventral lancets.

Now, contrary to popular belief, the poison-sac plays no part in the injection of the venom, this being effected by two pumps, consisting of two pouch-like lobes, attached to the upper edges of the two lancets and lying side by side within the cavity of the bulb of the sheath when the sting is retracted. When the worker bee has extruded her sting from the sting-chamber and jabbed it hard at her victim the tip of the sting just penetrates the skin and the two lancets, working alternately, begin sliding backwards and forwards along their tracks. As soon as one of them has just entered the skin it is held in place by its barbs whilst the other lancet slides forward and penetrates a little further than the first one. This process is repeated over and over again so that, by means of the rapidly alternating thrusts and pulls of the lancets, the complete sting is drawn deeper and deeper into the flesh. These same movements of the lancets also work the poison pumps, the plungers of which are the pouch-like lobes attached to their upper edges in the bulb chamber, and thus drive the venom down the sting canal until it escapes into the victim's tissues through a small slit near the tip of the sting. Usually long before the sting has had time fully to enter the flesh in this way the bee has broken free from her victim, leaving the sting and its poison glands behind, and the continued movements of the lancets drawing themselves into the flesh can readily be observed. The drop of venom introduced is, of course, very small and weighs less than half a milligram. Three systems of glands are said to be associated with the sting. One of these is supposed to produce a lubricant, whilst the other two, the paired, so-called, "acid glands" and the "alkaline gland" secrete fluids which together make up the venom when they become mixed in the canal of the sting when the bee uses it.

Unfortunately we do not yet fully understand the complex chemical composition of bee venom. The old idea that its chief constituent is formic acid (which supposition may have led to the idea of using ammonia and other alkalis as palliatives) is now known to be incorrect. All that can definitely be said is that bee venom is very complex, and not a relatively simple substance, and possesses a protein-like toxicity.

On an average person the result of a bee sting is the production, within the course of a few minutes, of a small, red, area at the point of injection of the venom, surrounded by a larger, whitish, area. At first sharp pain is felt at the point of stinging but this usually only lasts for a minute or so, and is followed a little later on by swelling, which may become fairly extensive, and later still by more or less irritation, heat and an itching sensation. The swelling and irritation vary in degree according not only to the part of the body that is stung, but also to individual idiosyncrasy. The first sharp pain at the site of the sting is said to be due to the presence of a trace of histamine, or some similar substance, produced by the acid glands, and the later toxic effects to something else in the venom.

After a beekeeper has been stung a number of times the reactions mentioned above gradually become less pronounced and no ill effects are usually noticeable. Many beekeepers find, however, that after a longish period, such as the winter, during which they have not been stung, the next few stings they receive once again produce a certain amount of swelling; but more or less complete immunity (depending upon the individual) is soon re-acquired.

Although the average person can quickly become relatively immune to the after effects of bee stings (e.g. the swelling) he always feels a painful sensation at the point of penetration immediately after being stung. The degree of pain experienced is usually very slight, but can be severe. It has been estimated that about two per cent of persons suffer from more serious effects when they are stung, but really severe results are very rare. In these very rare cases collapse and even death have been known to result from a single sting, presumably because the unfortunate person had unknowingly become hypersensitive to bee venom as a result of one or more stings received on some previous occasion. It should be emphasized, however, that such cases are *very* rare.

If a child is stung by a bee it is often helpful to apply something harmless such as blue-bag, onion or lemon juice, iodine, or even honey. It is extremely doubtful whether any of these serves any medicinal purpose, but they may have a useful psychological effect by distracting the patient's attention until the immediate pain is past. A cold water compress will often be found both soothing and useful in reducing any subsequent swelling. In cases of collapse, or severe shock, an injection of adrenalin is probably the most effective treatment. In such cases a doctor should, of course, be called immediately.

From very early times there appears to have been a widespread belief that repeated stinging by bees reduces the tendency of a beekeeper to suffer from rheumatism and kindred ailments, and many beekeepers today subscribe to this view. On the other hand there are well-known bee men who, although they undoubtedly receive plenty of stings during the course of a season, nevertheless suffer severely from rheumatism. Clearly this is a somewhat controversial subject and all that can, in my opinion, safely be said at the present time is that until we know considerably more about the nature of the various rheumatoid complaints, and also about the chemical composition and physiological action of bee venom, any forward steps that are made in the treatment of such complaints with bee venom will, necessarily, be the result of trial and error, rather than of work based on ascertained facts.

Those persons who have conducted experiments upon human beings suffering from rheumatoid complaints, and who support the view that treatment with bee venom is, in certain circumstances, effective, are themselves divided into two schools of thought. Some use living bees to inject the venom and claim that most, if not all, extracts of venom have lost their potency; others consider that only extracts of venom should be used, and say that only by the use of extracts can any standardised treatment be evolved.

There is no doubt that treatment with bee venom sometimes achieves spectacular results in the alleviation of rheumatism, but, alas, in many cases it fails in its object.

CHAPTER 12

COLONY ODOUR AND DEFENCE

*"In valour therefore, and magnanimity they surpass all creatures:
there is nothing so huge and mighty that they fear to set upon, and
when they have once begun they are invincible; - - so great hearts do
they carry in so little bodies."*
CHARLES BUTLER (1623). *The Feminine Monarchie. 2nd ed.*

MANY BEEKEEPERS, as a result of their observations, have been
led to suppose that every colony of honeybees possesses its
own particular odour which is common to all members of the
colony including the queen. Some have supposed that these
colony odours are inherited, others that they are acquired by
contact with the environment and are composite, being com-
pounded from the odours of the individual bees, of the brood,
combs, stores of food, hive materials, etc. Yet others have
supposed that colony odours are partly inherited and partly
acquired.

It has been further supposed that the sharing by all the
members of a colony of one particular odour, and the ability of
the individual bee to distinguish her colony odour from that of
a bee belonging to any other colony, is the means whereby bees
differentiate between friends and foes. Indeed even as long ago
as 1623, that great English beekeeper, Charles Butler, wrote :
" . . . and by this sense (the sense of smell) they find out any
strange bee, which is not otherwise to be known from their own
company, . . . ".

Until 1926 the idea that each colony of bees possesses its own
particular odour had, like so many other ideas, been based on the
accumulated experience of beekeepers rather than on adequate

experimental data. It has been known since the beginning of this century, however, that a worker honeybee that has been "lost" and has eventally succeeded in finding the entrance of her hive again will usually expose a small gland on the dorsal surface of her abdomen from which a highly volatile sweet-smelling scent is given off, and simultaneously, fan a current of air backwards over the exposed gland with her wings, thus distributing the scent (Pl. 14, p. 87). Other "lost" bees perceive this scent and are thus guided home and then they, in their turn, often pause at the hive entrance and expose their scent glands. Despite its apparently purposeful nature this action has been shown to be entirely automatic and so quite unreasoning and before 1926 no evidence was available to indicate that the scent given off by the bees of one colony differs from that produced by bees from other colonies. In the course of their work beekeepers have caused some of the bees of a colony to disperse scent near the entrance of their hive; and many have noted that the scent thus dispersed tends to attract to the hive bees from any colony in the apiary who happen to be flying nearby. Even though the existence of individual colony odours has been clearly established, there is considerable reason to suppose that the differences between the odours of bees belonging to different colonies are small, and that bee scent is very attractive to all bees irrespective of its origin.

In 1923 von Frisch reported his observation that when a foraging honeybee finds in the field a dish containing a plentiful supply of concentrated sugar syrup she will often expose her scent gland whilst feeding at it (Pl. 15a, p. 94) but that, in these circumstances, she does not fan with her wings and thus disperse the scent widely. He was able to demonstrate that exposure of the scent gland in this way serves to attract other bees to the dish. In 1926 von Frisch and Rösch described further observations in which they have found that some foragers, on returning to a dish for more syrup, expose their scent glands whilst flying around it preparatory to alighting. If the supply is plentiful and concen-trated, these foragers sometimes continue to do so for some time whilst feeding. They also reported that they had seen honeybees (*A. mellifera*) exposing their scent glands whilst feeding at flowers. Very few other people have reported seeing workers of *Apis mellifera* exposing their scent glands when visiting flowers, but, when visiting Ceylon, the author very frequently observed workers of *Apis indica* doing so. Furthermore workers of this species of honeybee often expose their scent glands in this way, without fanning, in and around their nests. Unlike the workers of *Apis mellifera*, they seldom seem to promote scent dispersal by fanning.

Von Frisch and Rösch were also able to demonstrate that whilst the scent produced by foragers flying around or feeding at a dish of syrup is attractive to all bees in the immediate vicinity of their colonies, it is even more attractive to other members of the scent-producing bees' own colony. This appears to be the first definite experimental evidence in support of the theory that each colony of honeybees possesses a scent which differs from that of other colonies, and which is distinguishable by the bees of the colony to which the scent-producing bees belong and preferred to those scents produced by bees of other colonies.

The importance of these observations of von Frisch and Rösch have now been realised and they have been confirmed and subsequently extended by Kalmus and Ribbands, who have shown not only that the scent produced at a dish of syrup by a number of bees can persist and still prove attractive to other bees for two hours or more, but also that each colony normally has its own particular scent which cannot be entirely hereditary. Indeed, they have produced evidence which, whilst by no means ruling out the possibility that heredity plays some part in determining the odour of a colony, suggests that colony odours depend to a considerable extent upon the food on which the members of each colony have recently been feeding. For example, if during the summer a colony of bees in dequeened and divided into three equal parts, each part being housed in a separate hive, one part being fed on pure sugar syrup and the other two parts on heather honey with its pronounced aroma, the bees of each part can be shown to have developed their own particular colony odours, within the course of a week or less. Those of the two parts fed on heather honey are apparently identical, and the members of these two parts will mingle together without hesitation and do not appear to be able to distinguish a bee from their own part from one from the other part. On the other hand bees from the syrup-fed part will not mix freely with bees from either of the parts fed on heather honey, and exhibit definite suspicion when they meet them in an arena. Bees from either of the parts fed on heather honey likewise exhibit suspicion if they meet a bee from the syrup-fed part. Since the bees in all three parts are genetically similar such differences in the body odours of the bees of the syrup-fed and heather-honey-fed parts, and hence of the colony odours of these parts, can only have been due to the acquisition of different scents from their environments, scents which could only have been due to differences in food supply. It must be remembered, however, that such changes in body odour, although probably brought about to some extent by the food eaten, could also be adequately explained by absorption of the scent of heather honey into the waxes with which bees' bodies are covered. It is known that such absorption of floral and other scents does take place.

Although in this experiment the final differences in the odours of the different parts (colonies) could not be explained by differences in heredity, there is evidence that such heredity differences do exist between colonies of different races of bees, and possibly (though probably to a less pronounced extent) between colonies of the different strains within any one race. Thus it is sometimes more difficult to introduce a queen of a light coloured race successfully to the bees of a colony of a dark coloured race, than to bees of a colony of her own light coloured race, and *vice versa*. Clearly, however, other explanations are possible.

It is probable, therefore, that the odour of any particular colony, and of each of its members, is made up of a hereditary component and of a number of other components which are dependent upon the environment at the time and may be either absorbed from it or taken together with food.

The odour of every colony of honeybees is thus continuously changing, as the food in use by each colony is continuously changing, in the proportions derived from the different kinds of flowers. No two colonies are likely to be using exactly the same food, except under very exceptional conditions such as might occur in certain places where one food plant only is present (*e.g.* a large heather moor). If more than one flower-source is available, the proportion of the foragers of each colony which is working on any particular kind will vary. This has been made clear by the results of pollen-trapping experiments. It has been found that if the hives of two colonies stand side by side, in a district where a number of different kinds of flowers are available, the proportions of different kinds of pollens brought into the hives during the course of a single day by the foragers are always different. The bees of one colony may, for example, be collecting mainly fruit-tree pollens and a little dandelion pollen, whilst the bees of another colony may be working almost exclusively on dandelion flowers.

It will be clear that a system of this kind presupposes that the members of each colony share a common food supply; that all the food brought into the hive by the foragers is, so to speak, pooled. That this does in fact occur has been demonstrated in a number of different ways. For instance, a few members of a colony have collected a small quantity of food in the field which was contaminated with a poisonous dust or spray and have returned home with it. Many (sometimes thousands) of the other members of the colony, both adults and larvae, have subsequently become poisoned and died, although the colony has had no further access to poisoned food. A number of years ago Park, in America, showed that when bees bring drinking water into their hive they share it out very quickly and widely between very many members of their colony. Nixon and Ribbands have also studied this phenomenon in detail using a radio-active tracer, and have shown how a very small quantity of sugar syrup (one table-spoonful) collected by half-a-dozen worker bees was very widely shared amongst the adult members of their colony and their larvae.

Specific colony odours, then, exist and are shared by all the members of the colonies. What part, if any, do such odours play in enabling the bees of a colony to recognise other members of their own colony, and distinguish them from members of other colonies? Have Charles Butler and many other beekeepers been correct in supposing that scent plays a fundamentally important role in colony defence? John Free and the author have investigated this question in some detail—and a related one: the part played by the behaviour of the bees themselves, when they approach the entrance of their own hive or that of another colony, in enabling the members of a colony to distinguish between friend and foe. Normally, during the active season, when enough fresh food to supply the colonies is being collected by the field bees, the colonies are quiet and peaceful, and few, if any, bees can be seen guarding the entrances of their hives. During the late summer and autumn (as well as at other times when the weather is warm enough for bees to be active but when relatively little food is coming into the hives) robber bees, and often wasps as well, frequently attempt to plunder the colony's stores of honey. A hive entrance then exhibits an entirely different picture. Numerous guard bees are present and fights often take place between them and would-be robber bees from other colonies (Pl. 28b, p. 155), and perhaps between the guards and wasps.

Guard bees can belong to almost any age group and some bees undertake guard duty before they become foragers. They busily intercept and examine all bees (including members of their own colonies) and wasps, which approach the entrances of their hives. Many guard bees are foragers as well, and alternate between foraging and guard duty (Pl. 29b, p. 162). On the whole the older and more thoroughly established foragers tend to leave guard duty to younger foragers. Each guard bee tends to patrol a certain limited area around the entrance of her hive, but will move elsewhere when necessary. Periodically a guard bee will leave her post and disappear, either within her hive or on a flight in the field. She may be away only for a minute or two or for several hours, and then reappear on guard duty once more in the same place as before. Just as the lengths of such off-duty periods vary enormously, so does the length of time spent on guard duty. During off-duty periods guard bees have been seen collecting both nectar and pollen in the field and even robbing the hives of other colonies of bees.

A bee on guard duty at the entrance of her hive (or on a comb which the beekeeper has exposed) assumes a very characteristic attitude. She stands with her front-legs off the ground and her mandibles and wings closed (Pl. 25a, p. 154) and frequently jerks round to watch any bee which she sees approaching. As soon as a bee alights, any guard nearby who has seen her do so attempts to intercept and examine her. Sometimes when a lot of intruders are about the guards become very excited and wait with open mandibles and wings all ready to dart towards any bee which alights near them (Pl. 25b, p. 154).

Before we consider the mutual reactions between guard bees and the bees which they succeed in intercepting, we must first discuss how the peaceful, unalerted, colony with no guards on duty becomes alerted in the first instance.

When colonies of bees are gathering enough fresh food for their immediate requirements and are unalerted, with no guards at the entrances of their hives, we may quietly interchange the positions of pairs of occupied hives. Any foragers that are in the field tend to return to those positions in which their hives have previously been standing. If the hives are moved gently and not jarred their colonies remain unalerted. The mutual reactions between the owners of a hive, and the strange bees from other colonies which "drift" inadvertently to it, mistaking it for their own, may then be seen. Is the presence of strange bees in and around a hive enough to alert its owners? We find that those bees which have been gathering pollen and nectar, and have drifted to strange hives, enter them without hesitation and without being examined or molested in any way by the owners of the hives. It is clear, therefore, that strange colony odours do not deter bees from entering the hives of strange colonies to which they have drifted. But if the hives whose positions have been interchanged are thumped vigorously instead of being moved as quietly as possible, the inmates are greatly disturbed and guard bees quickly appear at the hive entrances and attempt to intercept and examine all bees which try to enter. The disturbance has had the effect of alerting the colonies. Nevertheless, most of the nectar- and pollen-carrying bees which drift inadvertently to hives other than their own continue to enter them without hesitation and in the great majority of cases without difficulty. These laden foragers are often intercepted by guards who examine them with their antennae (which bear their olfactory organs); but the intruders usually pay no attention and behave in a dominant fashion, running into the hives, with the guards running after them in some cases and still trying to examine them. The guards usually make no attempts to seize them. Dominant intruders of this kind continue to behave inside the hive as though it were their own, getting rid of the food they have collected and even dancing and recruiting other bees as foragers. If they do notice a strange hive odour, or the strange body odours of their companions, it does not appear to cause them to alter their behaviour in any way. On the other hand, it seems likely from the way in which they sometimes follow and attempt to examine them, that the guards are often doubtful about the identity of these intruders.

Some of the bees which drift to the hives of strange, alerted, colonies are not, however, carrying loads of pollen or nectar and many of these when intercepted by guards are examined for a few seconds only before being subjected to mauling, the guards pulling them away from the hive entrance by their legs and wings.

Although colonies can be alerted by thumping on their hives, such violent disturbance is seldom likely to occur; and two other things which cause colonies to become alert have been discovered. If large numbers of bees drift into the hive of a strange colony some of the owners of the hive sooner or later become suspicious, probably because of the intruders' strange odours, and begin to intercept and examine them. In fact, guard bees make their appearance first within the hive and then at the entrance. It is also possible to alert colonies of bees by mimicking the character-istic flight behaviour of robber bees, by bobbing in front of the hive entrances dead bees tied on the ends of pieces of cotton. It does not appear to make any difference whether the dead bees used are members of the experimental colony or not, thus indi-cating that it is the bobbing movements of the dead bees, rather than any scent which they possess, which causes the bees to regard them as intruders. Lecomte has shown that guard bees will even attack models of bees, made out of cardboard and other materials, when these are moved in front of them. Guard bees appear to have an inherent urge to attack any object which is swaying or jerking about in front of the hive entrance.

We can conclude, therefore, that apart from such stimulation as violent vibration of their hives, colonies of bees become naturally alerted either by the presence of robber bees (or wasps), with their characteristic swaying flight (which they make before the hive entrance whilst they appear to look for opportunities to slip into the hive unobserved by the guards) or by the presence of numbers of strange bees which have drifted inadvertently to the hive. Once alert, a colony may remain so for only a few minutes, or for days on end. Colonies of different strains of bees vary greatly in this respect, as many beekeepers have good reason to know; for it is the alert guard bees which attack them. The length of time during which alertness persists also varies, even in the same colony, at different times of the year. When nectar is scarce colonies are alerted easily, during major "honey-flows" only with difficulty. During "honey-flows" the bees of the colony become so preoccupied with the collection of food that they largely, perhaps entirely, ignore even large numbers of strangers in their midst. On the other hand, at those times when colonies can readily be alerted, the guards very occasionally become so excited that they are liable to attack any strange bees which come near their hives.

Some years ago Kalmus fed bees at a dish of syrup. When the dish contained plenty of syrup, and there was plenty of room, bees from two or more colonies took little notice of each other (Pl. 15a, p. 94). But when the supply of syrup in the dish began to fail the bees of the different colonies began to squabble and attack one another with their mandibles (Pl. 15b, p. 94). These quarrels never occurred between bees of the same colony. Almost certainly bees distinguish between members of their own colonies and bees from other colonies by differences in their body odours.

When a guard bee intercepts a member of her own colony, or any intruder other than a robber bee, she examines it with her antennae. Within a matter of seconds she identifies, by means of their body odours, the majority of the bees she intercepts, either as members of her own colony or as intruders from other colonies. Sometimes, however, the guards appear to be doubtful about the identity of a bee and each will examine her body thoroughly for half-a-minute or more at a time. Bees identified as members of the guards' own colonies are allowed to pass without delay, but intruders, other than dominant ones, are stopped and their subsequent fate depends on how they behave.

Robber bees are recognised by the guards by their characteristic, hesitant flight before the hive entrance. As soon as a robber alights, any guard bee who has seen her do so immediately attempts to seize her and sting her to death without any preliminary examination. The robber bee always attempts to escape from the guard bee, but, should she be unable to do so, she will seize the guard bee in her turn and attempt to sting her. These fights are very violent as each bee attempts to sting the other, and the contestants often lie on their sides and spin round and round like a catherine-wheel as they grapple with each other and each attempts to find a place in her opponent's body into which she can thrust her sting (Pl. 28b, p. 155). Sometimes the guard bee and sometimes the robber is stung, and the victim crawls away and dies within a few minutes. The victor is unhurt and has no difficulty in withdrawing her sting from the body of her victim, but sometimes experiences difficulty in breaking away from her clutches.

Dr. C. R. Ribbands expressed the opinion that it is unlikely that robber bees have evolved a pattern of behaviour to their own disadvantage, and he has suggested that their characteristic flight is a consequence of their recognition: that, indeed, robber bees as well as all other intruders are recognised as such by their alien body odours. This does not, however, appear to be the correct interpretation of this behaviour, as Free and the author have been able to show that every bee has an innate tendency, when robbing a hive, to fly in this manner whenever she sees a crowd of bees near its entrance. It makes no difference whether or not the bees forming this crowd belong to the would-be robber's own colony and have or have not the same body odour. Indeed, it has been possible to arrange an experiment in such a way that marked robber bees were induced to attempt to rob a hive containing about half the members of their own colony with its brood, queen, etc., which had been taken from the parent hive a few minutes earlier. On seeing the guards (members of their own colony) at the entrance, these bees exhibited the characteristic, hesitant, flight of the would-be robber, and when they alighted and were intercepted they immediately attempted to escape. However, some of them were seized and stung to death. The only alternative to the explanation that they were identified as robbers by their behaviour, seems to be the hypothesis that robber bees produce a "fear" odour and are identified by it. Such a hypothesis seems quite unnecessary and does not explain why bees will attack an entirely artificial bee-like lure when it is dangled in front of their hives.

Inadvertent intruders which are not carrying pollen or nectar adopt a submissive attitude towards the guard bees who intercept them, making no attempts to run away and turning their extremities away from the guards and tucking in their tails (Pl. 26a, p. 154). The guards move round the bodies of such submissive intruders examining them with their antennae and, when a guard approaches the head of such an intruder, the intruder turns towards her and offers her a drop of regurgitated food between her open mandibles (Pl. 26b, p. 154). This offer of food is usually refused. Even if it is accepted this does not seem to help the intruder in any way. When her offer of food has been refused several times in succession, the submissive intruder suddenly extends her tongue and begins to "strop" it between her front-legs (Pl. 27a, p. 154). Such tongue-stropping differs from normal tongue-cleaning in several important respects. It is, for example, carried out more rapidly and also for a longer period of time than normal tongue-cleaning. It is an example of a "displacement activity", and is probably induced in this instance by frustration of the intruder's two urges, to enter the hive and to give food.

Once the guards have identified a submissive bee as an intruder they proceed to maul her, pulling at her legs, wings and sometimes even her mouthparts with their mandibles, and dragging her away from the hive entrance. A weak individual, such as a very young bee, may even be picked up bodily by a guard bee, who will fly off with her and drop her unhurt some distance away from the hive.

In exceptional circumstances members of the guard bee's own colony are treated as intruders and are mauled and dragged away from the entrance of their own hive. This occurs when the submissive bees are suffering from a complaint, a form of bee paralysis, which, it is believed, causes changes in their body odours so that they are no longer recognised as members of their own colony.

Submissive intruders are never stung unless they try to escape from the guards, or appear to be trying to do so, as can happen when they are being pulled about a great deal by excited guards. The submissive state of the intruder appears to inhibit the guards from stinging her. Any attempt to escape on the part of an intruder immediately releases the stinging response of the guards; they usually climb on top of the intruder and attempt to drive their stings into her body (Pl. 28a, p. 155). Occasionally the stinging response of a guard bee can be evoked experimentally by holding a submissive intruder with forceps by a wing and jerking her away from the guard bees.

The stinging responses of guard bees are, therefore, always released by the darting behaviour of robber bees, and may be released by the sight of inadvertent intruders, who have hitherto been submissive, trying to escape. Also, if excited guard bees of two colonies meet, as sometimes happens when beekeepers attempt to unite two colonies together without taking sufficient care, severe fighting often occurs.

On rare occasions the legs of a submissive intruder who is being mauled are all jerked quickly and simultaneously off the ground, in which case she will equally suddenly adopt an entirely passive state (a condition of thanatosis) and allow herself to be carried around by the guard bees (Pl. 27b, p. 154). On being released intruders which have assumed this passive state clean themselves up and usually continue on their way towards the hive entrance apparently quite unhurt.*

In spite of the activities of the guard bees some intruders do succeed in entering hives, but they are liable to be subjected to exactly the same treatment on the combs within the hive as they would be on its alighting-board (Pl. 29, p. 162). Many of them are mauled and dragged out of the hive again. However, some of them usually do succeed in remaining within the hive, and, after about a couple of hours, they cease to be further examined and mauled and become adopted as members of the colony. Presumably during the two hours they have been in the hive these

---

*This state of thanatosis, which is adopted involuntarily by the submissive bee, should not be confused with the passive attitude with outstretched wings which worker bees adopt voluntarily when their bodies are being cleaned by other bees.

bees acquire similar body odours to those of the rightful occupants of the hive. These are probably acquired by the absorption of the general hive odour into the waxes on the surfaces of their bodies and by contact with the bees, and, perhaps, to some extent by sharing in the communal food supply. The intruders are, therefore, no longer recognisable as strangers. Such adopted intruders have been seen, twenty-four hours after they first entered a strange hive, guarding its entrance against intruding members of their parent colony.

During a visit to Ceylon the author was able to observe the behaviour of workers of both *Apis indica* and *Apis florea*. He found that intruders and guard bees of these species exhibit behaviour patterns which appear to be identical with those exhibited by *Apis mellifera*. Workers of these two species and also of *Apis dorsata* also exhibit the curious "shimmering" behaviour (p. 14) when disturbed by intruders such as hornets or men.

We can conclude, therefore, that body odours play a fundamental part in enabling honeybees to distinguish between members of their own colony and intruders from other colonies. It is also clear that the behaviour patterns of intruders are extremely important, as they not only enable guard bees to recognise robber bees very quickly, but also serve to preserve the lives of most of those intruders who will not harm the colonies whose hives they are seeking to enter.

CHAPTER 13

COLONY REPRODUCTION OR SWARMING

"The stocks having bred and filled their hives do send forth swarms. A swarm doth consist of all such parts as the stock doth: namely of a queen bee, honeybees as well old as young, and drone bees."
CHARLES BUTLER (1623). *The Feminine Monarchie. 2nd ed.*

COLONIES of all species of honeybees swarm. Swarming is the method by which colony reproduction, as distinct from the reproduction of individual honeybees, takes place. Just as reproduction of individual bees within a colony is essential to the continued existence of the colony, so reproduction of colonies is essential to the continued existence of the species, for without it any increase in the number of colonies would be impossible as would replacement of any colonies destroyed by disease, or other causes. Swarming is an inherent characteristic of honeybees; and it is most improbable that nonswarming (in the literal sense of the word) strains of honeybees will ever be developed. It should at the same time be realised that colonies of some races of honeybees (and indeed colonies of some strains of the various races) swarm much more frequently than others: the practicability of producing strains of bees in which the swarming tendency is exceptionally low has been convincingly demonstrated by a number of modern selective breeders.

The desire of many beekeepers to reduce the incidence of swarming amongst their colonies to a low level is only a recent development in the long history of beekeeping. Little more than fifty years ago swarms, at any rate those that were obtained early in the season, were regarded as highly desirable by the great majority of beekeepers and played an important part in their systems of management. Hence such rhymes as: "A swarm of bees in May, is worth a load of hay; a swarm of bees in June, is worth a silver spoon". Today swarming bees still possess the power to fascinate people, including experienced beekeepers, as much as they ever did; but swarming is regarded as undesirable by many beekeepers. They consider that the issue of a swarm from a colony will be accompanied by a reduction in the size of the honey crop which would otherwise have been obtained by this colony. Hence many different methods of so-called "swarm control" have been elaborated, most of which aim to thwart this natural tendency of honeybees.

We know very little about the stimuli which cause bees to swarm. At least two theories, the "brood-food" theory and the "over-crowding or congestion" theory, can explain many of the known facts adequately; but neither they, nor any other theory which appears to have been put forward so far, explain all the facts. However, before theories we must consider facts.

The swarming of honeybees must not be confused with the so-called swarming habits of ants and termites, whose "swarms" are composed of the royal forms—the drones and virgin queens—on their nuptial flights. During these flights the royal forms seek out one another; those of ants pair immediately, of termites later on. After the nuptial flight each royal pair of termites, or impregnated ant queen, seek to found new colonies entirely on their own, in very much the same sort of way as overwintered, impregnated, queen bumblebees and the queens of our social wasps do in spring.

The honeybee swarm, however, consists not only of a queen (sometimes more than one queen—in which cases all will almost certainly be virgins) but also of a large number of worker bees and, very often, of a few drones as well. The queen may or may not have been impregnated. If a swarm contains an impregnated queen it will always seek a new home of its own, whereas if it contains one or more virgin queens it may either seek a new home or may return to the one that it has just left. In the latter case the workers and drones have usually accompanied a virgin queen (or queens) on her mating flight and this type of swarm is known as a "mating" swarm.

Another type of honeybee swarm is sometimes recognisable, the type which in Europe is usually known as a "hunger" swarm, and which occurs when a colony is starving. In such cases the colonies usually have either no unsealed brood or only very little. As Charles Butler observed early in the seventeenth century, the bees will often drag out of their cells the larvae and pupae, particularly those of the drones, when there is little food in the hive. Apart, perhaps, from a dozen or two stragglers who get left behind, *all* the bees leave the hive together and seek their fortune elsewhere. This type of swarming occurs very frequently among the Eastern honeybees (*Apis indica*), Giant honeybees (*Apis dorsata*) and Little honeybees (*Apis flora*), but very much more rarely among Western honeybees (*Apis mellifera*). This type of swarming of these eastern species of honeybees is usually known as "migrating" or perhaps, more accurately, as "absconding". It has been suggested by Ribbands and others that that type of swarming of the Western honeybee (*Apis mellifera*) which results in colony reproduction has been derived from this absconding type of swarming. However, this is improbable, as when a colony absconds all the bees quit their old home and colony reproduction never takes place. Absconding appears to serve one purpose only, that of removing the *whole* colony from an unfavourable environment and giving it a chance of finding a more favourable one. Colonies of *Apis indica*, *A. florea* and *A. dorsata* abscond from their nests not only when they are on the verge of starvation, but also when their nests are invaded by wax-moths, ants or other enemies, or even sometimes when they are disturbed by man, as beekeepers in India and Ceylon know only too well. But colony reproduction occurs in *Apis indica*, *A. florea* and *A. dorsata* by exactly the same type of "reproductive swarming" as it does in the case of *Apis mellifera*: a mated or virgin queen leaves the parental nest together with a *proportion* of the worker bees and seeks to set up an entirely new colony in a new home, leaving the old, parental, colony to carry on.

It seems much more probable that the type of swarming which results in colony reproduction has evolved from so-called "mating" swarming than from "absconding" swarming.

Mating swarms occur frequently among *Apis indica* (and

Plate 25.—GUARDING OF THE NEST (I)
above. Pair of guard bees at hive entrance (x 3·0).   *below.* Highly excited guard
bee (x 3·0)

Plate 26.—GUARDING OF THE NEST (II)
above. A submissive intruder turns her extremities away from an examining
guard bee (x 3.3). *below.* A submissive intruder offers food to a guard bee. In
this case it is being accepted (x 2·8)

Plate 27.—GUARDING OF THE NEST (III)

above. 'Displacement' tongue-stropping. This differs from normal tongue-cleaning and is probably a sign of frustration (x 2·8). *below.* The intruder has adopted an entirely passive state—a state of 'thanatosis.' She is quite unhurt and is carried away from the hive entrance by the guard bees (x 2·8)

Plate 28.—Guarding of the Nest (IV)
above. Stinging response of guard bee released by attempted escape of intruder who is being severely mauled and dragged away from the hive entrance (x 3.0).
below. Fighting to the death between guard bee and robber bee. Two bees fighting in this way usually seize one another and lie on their sides, head to tail, attempting to sting each other (x 3.0)

probably *A. florea* and *A. dorsata*), and much less frequently among *A. mellifera*. When produced by a very small colony they sometimes consist of virtually all its bees; but those from larger colonies* often consist only of a proportion of the adult population, and in this way are more comparable to that type of swarm which results in colony reproduction. Although mating swarms frequently return to the hives which they have just left, some fail to do so and set up new homes on their own—this is particularly true of those from larger colonies which are similar in composition to "casts" or "after-swarms". An "after-swarm" leaves the hive with a virgin queen, after the "prime" swarm with the old, mated, queen has left and established a new home of its own**. It is often similar in behaviour in that it attempts to set up a new home of its own. Thus mating swarms sometimes result in colony reproduction which the absconding of colonies never does.

When the so-called "swarming impulse" of a colony of honey-bees has been aroused, the worker bees proceed to construct a number of special cells, the queen cells. These are small at first and shaped rather like tiny inverted thimbles with their mouths downwards, and are often attached to the sides or bottoms of the combs (Pl. 12a, p. 83). A certain number of these "queen cell cups" are present in almost every hive in summer, but when a colony is making preparations to swarm their number is increased. The queen lays eggs in some of these cell cups, perhaps in a dozen or more. Some observers claim that the bees induce her to do so, but it seems more probable that she does so quite automatically and without the necessity of any special inducement, simply because they are clean and sited in the brood area (p. 68). The worker bees feed the resulting larvae with brood-food secreted by their hypopharyngeal and mandibular glands and the cell walls are quickly enlarged to accommodate the rapidly growing

*Such a mating swarm is sometimes induced by a beekeeper if he allows a large colony to rear its own queen. Some large colonies not only lose a proportion of their bees in such mating swarms but are also rendered hopelessly queenless.

**This is true of most swarms because they come from overcrowded colonies. An uncrowded colony seldom swarms; when one does, the first queen that is reared usually kills the old, mother queen, and then leaves with the swarm.

larvae and their normally very abundant food supplies (Pl. 12b and c, p. 83). Meanwhile the old mated queen has been producing progressively fewer eggs each day, and may indeed have stopped laying altogether. Her abdomen has shrunk considerably, as it no longer contains a large number of fully formed and almost full formed eggs. Soon after the first of the "swarm queen" cells has been sealed, the old queen* leaves the hive accompanied by many of the worker bees of the colony (mostly young and middle-aged ones) and perhaps a few drones. After the whole lot have flown around the apiary for a short time they usually settle on a limb of a tree, or other convenient support near their old home and cluster together (Pl. 30, p. 163). After a time, which may vary from less than an hour to a couple of days, the whole swarm takes to the air once more and moves off to occupy a cavity in a hollow tree or other suitable nesting site. This has been discovered by scout bees, who sometimes clean it out to a certain extent. Very occasionally the bees of a swarm of *Apis mellifera* or *A. indica* appear to be unable to find a suitable nesting site, and build a nest in the open (Pl. I, frontispiece). Exposed nests of this kind are rare and seldom survive for very long, as they are attacked (particularly in winter when the bees are less able to defend themselves) by mice, etc. and are also very liable to be damaged by wind and rain. A swarm of this kind that contains a mated, laying, queen, the mother of a colony, is usually known as a "prime" or "first" swarm.

Usually about eight days after the prime swarm has left, one of the virgin queens emerges from her cell in the parent hive and, unless the worker bees prevent her from doing so, seeks out the other virgin queens in their cells and destroys them. Although the virgin queen often begins to tear open cells containing potential rivals, the worker bees usually complete the destruction; and there is some doubt whether it is in fact the virgin queen or the worker bees who kill the unfortunate inmates of these cells.

Although the evidence is weak, some people believe that a virgin queen "pipes" (p. 62) when seeking out rivals and, that queens still in their cells reply, so aiding their own discovery and destruction. A day or two after her rivals have been destroyed, usually when she is five days old, the virgin queen leaves the hive

*See second footnote p. 155.

to mate, either on her own or in a mating swarm (p. 153). If she returns successfully, as she usually does, she becomes the new, laying queen of the colony. Bad weather may postpone her mating; if it lasts for three or four weeks it may prevent it altogether, and the queen may become a "drone layer", laying only unfertilised, drone-producing, eggs: her colony is doomed to extinction unless she is replaced.

Sometimes after the first swarm has left, the worker bees keep the first virgin queen to emerge away from the other queen cells. This they do apparently deliberately, and they may even keep the other virgins in their cells for as long as several hours by clustering on top of them. In such cases, of course, the workers feed the imprisoned queens. If two virgins happen to emerge and meet one another a fight to the death occurs between them, but they do not usually meet in this way. Usually the first virgin leaves the hive with another swarm of bees, which is known as a "cast" or "after swarm". Several such casts may leave the parental hive one after another, each with a virgin queen; and two or more casts will sometimes join together, in which case the virgins fight to kill one another until only one remains to head the new colony. Swarms containing mated queens will sometimes join together too, and in such cases several queens may continue to live in the one, multiple colony for some time. Eventually one virgin queen remains to mate and subsequently head the parent colony, which may have become very depleted of bees after losing a prime swarm followed by several casts.

Sometimes the worker bees of a colony will tear down queen cells which they have started, even those that are in an advanced stage of development. Probably this is caused by some change in the environment, perhaps a weather change accompanied by a change in foraging conditions: many more data are required for a definite statement on this point.

The construction of swarm queen cells is almost always associated with a reduction in the colony's foraging activity and a reduction in egg-laying and hence in brood-rearing. This inactivity is accompanied by increased development of the brood-food (hypopharyngeal) glands of the workers and, possibly, development of their ovaries. Many beekeepers believe that preparations for swarming are triggered off by a change in the weather affecting foraging conditions. The sudden onset of a period of bad weather, during which the bees are confined to the hive, immediately after a period during which foraging has been good, often appears to be associated with swarming. At other times the reverse seems to be true, and colonies will start queen cells immediately good foraging conditions return after a period of bad weather and confinement to the hive. But should the weather become poor again they will tear the queen cells down. As we shall see presently, both of these apparently contradictory conditions, first, confinement of the field bees to the hive by bad weather, and, secondly, a rapid increase in the quantity of food stored within the hive as a result of good foraging conditions, can result in over-crowding of the occupied, and immediately occupiable, space in the hive and trigger off the swarming impulse.

It used to be thought that the queen always left the hive at the head of a swarm with the bees following her, but it is now known that this is by no means the case. The queen leaves the hive amongst the other bees, often after many, or even most, of them are actually in the air, and sometimes seems to be forced by the bees to leave. Often a swarm will leave a hive, and even cluster for a time, only to return home again because the queen has failed to join them. Only in a genuine mating swarm does the queen appear to be among the first to leave the hive.

One authority on beekeeping has stated that queen supersedure (p. 105) occurs most frequently towards or at the beginning of a honey-flow and that colonies frequently combine supersedure with swarming. Indeed, he has expressed the opinion that more swarms result directly from queen supersedure than from anything else. He has pointed out that "supersedure" swarms leave the hive with one or more virgin queens and that the old, mated, queen (the mother of the colony) remains in the parent hive and is replaced later on by a young, newly mated queen, one of her own daughters. No doubt it is true that queen supersedure, which is due as we have seen, to an insufficiency of "queen substance" (p. 105), is often the direct cause of the emergence of a swarm from a hive. It seems probable, however, that such a swarm is merely a form of mating swarm which may or may not return to the parent hive. Several such swarms may issue separately from the same hive and only one return with a mated queen, who may live quite happily alongside her mother for a while but eventually supersedes her as head of the colony. Although it is very probable that a deficiency of "queen substance" plays a direct and most important part in the development of both the supersedure impulse and the swarming impulse, these two impulses almost certainly have quite different origins and should not be confused with one another.

In a genuine case of supersedure the factor which results in a reduction in the amount of "queen substance" available to the bees is almost certainly the ageing or illness of the queen (Chapter 8, p. 105). It becomes necessary that she be replaced if the colony itself is to continue to survive. In a case of reproductive swarming, however, the survival of the colony is not at stake, although the survival of the race does depend upon this method of colony reproduction. In this case, although toleration of swarm queen cells containing eggs or larvae is probably due to the worker bees failing to obtain sufficient "queen substance" to satisfy their requirements, this lack is certainly not due to ill-health of the queen. Occasionally a mated queen who leaves with a swarm will head a large colony successfully for a year or two afterwards. This insufficiency of "queen substance" may be caused, as we shall see, in one of several possible ways and it has become clear that in the case of a colony that is reproducing itself by swarming the shortage that occurs may only be temporary. In such a case, once the queen is re-established in her new home she appears to be able to produce enough "queen substance" to satisfy her colony's needs once more.

Before we consider any further the probable part played by "queen substance" in swarming, let us first consider the two chief theories of the case of swarming, the "Gerstung" or "Brood-Food" theory and the "Demuth" or "Over-crowding" theory. Very briefly the Gerstung hypothesis supposes that if in a colony the number of nurse bees with fully functional hypopharyngeal glands rises to such a level that more brood-food is being produced than is required to feed the larvae that are present, then the nurse bees are eventually compelled to find an outlet for this surplus brood-food by rearing and feeding queen larvae who require very large quantities of it.

Gerstung clearly believed that most, if not all, worker bees pass through a stage when their brood-food glands secrete abundantly, and that these bees have to find an outlet for this brood-food by feeding larvae or the queen. He supposed that this resulted in the nurse bees at first feeding the queen lavishly with some of their surplus brood-food which, he believed, caused her to lay even more eggs than before, from which came more larvae requiring brood-food, thus relieving conditions for the time being. A little later on, however, these larvae developed into nurse bees themselves and the total surplus of nurse bees, and hence of brood-food, over that which was required to feed all the larvae in the hive, became worse than ever. He further supposed that the capacity of the queen to absorb the surplus brood-food was limited and, a new outlet for it was found in the shape of drone larvae which were then reared. However, even this usually proved insufficient to dispose of the surplus brood-food and swarm queen cells were started in order that the bees might relieve themselves of their unwanted surplus of brood-food by feeding queen larvae lavishly with it, so eventually swarming resulted.

During the 1930's Morland reviewed and elaborated Gerstung's theory, and it is probably still the most widely held theory of the cause of swarming in Britain today. It certainly explains many of the facts that are known about swarming, and has received tremendous support from the apparent success of several methods of swarm control, such as the Snelgrove system, which have been based upon it. But Gerstung was incorrect in supposing that every worker bee becomes a nurse bee at some stage in her life and that there is no way in which the balance between nurse bees and larvae requiring to be fed can be redressed when necessary. Actually bees can forgo hive duties, including nursing, almost entirely when necessary and become foragers when only a few days old. Neither his theory, nor any modification of it of which the author is aware, is able to explain all the known facts. For example, why is it that a colony which continually has a great deal of sealed and emerging worker brood added to it, does not by any means always make preparations to swarm? This brood must give rise, according to the Brood-Food theory, to a tremendous surplus of nurse bees and brood-food. Why, again, is it that when a colony is "artificially swarmed" by shaking all the bees on to empty, broodless, combs, the surplus of brood-food over requirements for larval feeding, which must undoubtedly suddenly ensue, does not result in swarming?

According to the Demuth or Overcrowding (Congestion) theory, the only factor which is common to all colonies in which swarm queen cells are found is congestion, or overcrowding, in that part of the hive which the bees are able to occupy—this does not necessarily mean the whole of the interior of the hive. On some occasions bees will swarm when they are by no means occupying the whole of their hive. For instance, if the brood area of the colony is flanked on either side by undrawn sheets of wax comb-foundation and, for one reason or another the bees do not draw these sheets out into comb, congestion may occur in the brood area and swarming ensue. Idle foragers confined to the hive by bad weather can promote such congestion, as can insufficient storage space. Overheating due either to the placing of the hive in a position in which it receives full sunshine every day, or perhaps to insufficient ventilation, is only one of many other factors which have been supposed by various people to tend to promote swarming.

Now there does not appear to be any doubt at all that overcrowding of the hive space often, *but not always*, results in swarming. This may be caused by the hive being too small for the size of the colony, or equally well by some barrier (such as a comb of honey or a sheet of foundation) which prevents the bees from occupying the necessary space, even when it is available. In some experiments on swarming which one of my colleagues, James Simpson, made several years ago, a number of colonies had much sealed brood, and hence young bees, added to them. The degree of overcrowding became so great, in many cases, that large numbers of bees were unable even to get inside their hives; and hung clustering together outside their entrances day and night. Such overcrowding as this is much greater than any which occurs naturally, and yet not all of these colonies began preparations to swarm.

We must conclude, therefore, that neither the Brood-Food Theory nor the Congestion theory are capable of explaining adequately all of the facts that are known about swarming. We do not know yet the cause of swarming. However, let us return to a consideration of the part which "queen substance" may play in swarming behaviour, as it seems possible that it may be a clue to the problem.

We have already seen (Chapter 8, p. 97) that there is good evidence in support of the theory (which I have propounded as a result of my experiments during recent years and which has become known as the Queen Substance theory) that a mated, laying, queen honeybee possesses on all parts of her body substances which some of the younger bees of her colony appear to seek. Provided that the worker bees of a colony obtain enough of this substance from their queen they remain inhibited from either raising or tolerating the presence of queen larvae in their nest. The production both of emergency queen cells and of swarm queen cells is thus inhibited, and neither supersedure nor swarming can take place. But if for any reason the supply of "queen substance" falls below a certain minimum threshold level, this inhibition is removed and either emergency queen cells are built from cells containing young worker larvae, or queen larvae in swarm queen cell cups are both tolerated and reared. In other words swarming becomes possible. Furthermore "queen substance" acts in a quantitative manner, and the smallest amount required to inhibit a given number of worker bees from either building emergency queen cells, or tolerating the presence of queen larvae in queen cell cups, is not enough to inhibit a larger number of bees of the same strain from so doing.

In cases of supersedure a reduction in the quantity of "queen substance" below the threshold for inhibition of emergency queen cell production, appears to be due to some defect in the queen, probably one caused by old age or disease. Abundant evidence

---

Plate 29 (opposite). GUARDING OF THE NEST (v)
Some involuntary intruders do succeed in entering the nest. There they are subjected to further examination and many are dragged out of the hive again. However, some do succeed in remaining and eventually become adopted as members of the colony.

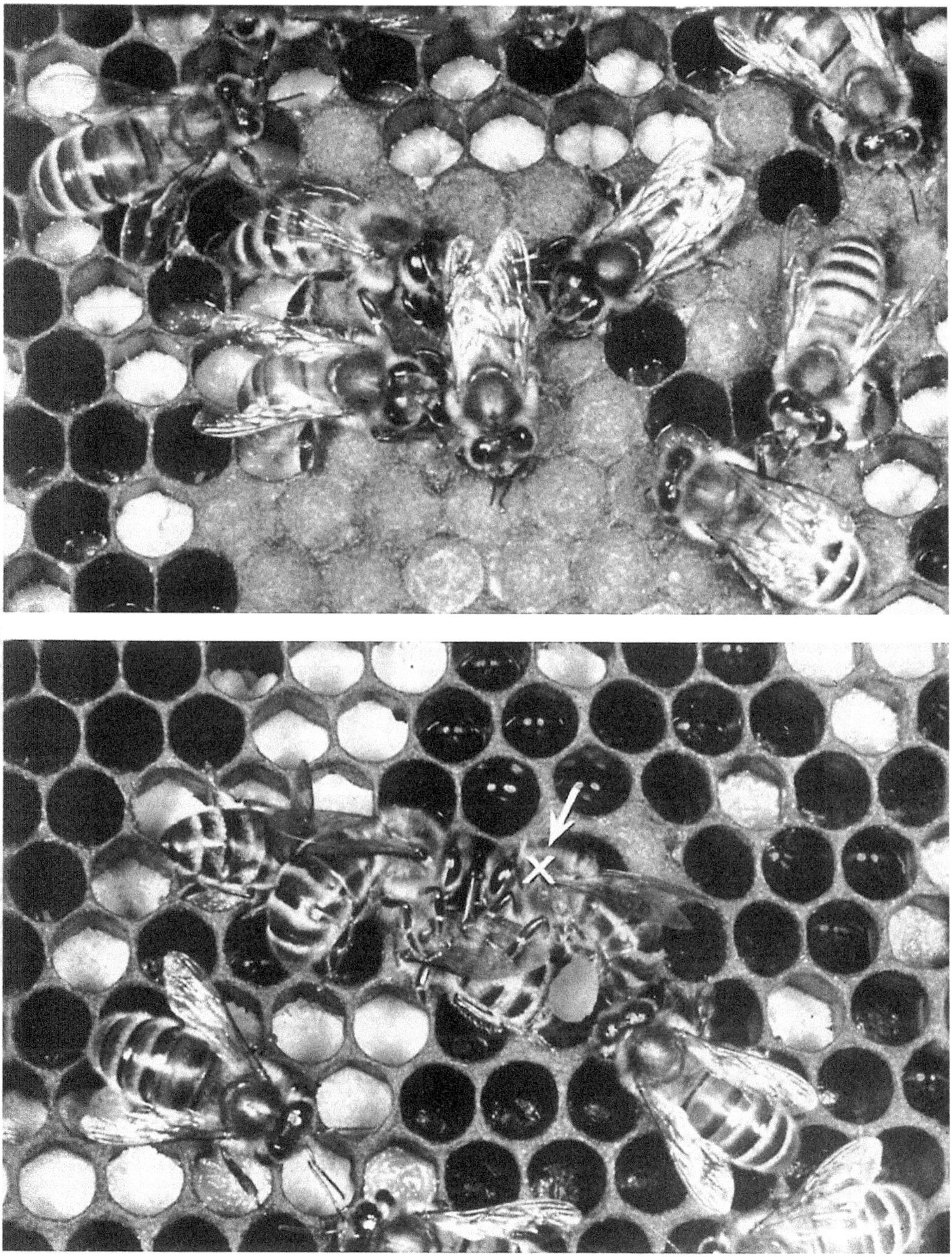

Plate 29.—GUARDING THE NEST (V)

above. Intruder being submissive and being examined on comb of brood in hive
(x 2.4). *below.* Submissive intruder being examined on comb of stored pollen.
Note bee (x) with pollen in pollen-baskets acting as guard and examining the
intruder (x 2.4)

Plate 30.—LARGE SWARM OF HONEYBEES ON BROOM BUSH
This swarm was about three feet long

has been obtained that any reduction in the amount of "queen substance" produced is not directly connected with any reduction in the number of eggs laid; although a reduction in egg laying usually occurs, and may be a result of the same factor that is causing a reduction in the quantity of "queen substance" being produced.

It seems very probable that when bees swarm something unknown causes a reduction in the quantity of this inhibitory substance which is being produced by the queen of the colony or, alternatively, that the quantity of inhibitory substance being produced remains more or less constant, but that its collection by and distribution among the worker bees somehow fails.

In each case an insufficiency of "queen substance" would lead either to the construction of emergency queen cells or to the toleration of eggs and larvae in swarm queen cells.

The second of these possibilities seems, on the whole, to be the more probable of the two. But although it seems pretty clear that the amount of "queen substance" available does become insufficient to inhibit the bees from rearing queens, we do not know to what such a reduction is due. Possibly some factor, such as unfavourable foraging conditions, leads to overcrowding of the brood area by adult bees and results in jostling and disturbance of the queen, which in turn somehow affects the production of "queen substance" or the amount of it which the bees are able to get from the body of their agitated queen. A Russian worker, Taranov, has indeed described how the nest of a colony which is beginning preparations to swarm becomes overcrowded with adult bees and how the queen is mobbed and jostled. Taranov thinks these are surplus nurse bees, but this now appears unlikely. He has suggested that the excess nurse bees cluster together and become inactive "swarm" bees with no work to do until they eventually leave the hive with the swarm.

A number of people, who have observed the behaviour of a queen honeybee with her workers, have noted that the foragers of a colony appear to pay no attention to their queen. They neither feed, groom nor examine her. Furthermore, during the course of the author's own experiments evidence was obtained which strongly suggests that established foragers are not interested in "queen substance" and do not seek it; and that, whereas household bees always attempt to build either emergency, or swarm, queen cells in the absence of enough "queen substance", well-established foragers do not do so. For example, an experimentally produced colony, consisting entirely of foraging bees and some brood of all ages but no queen, takes very much longer to produce emergency queen cells than does either a queenless, but otherwise normal, colony or one composed entirely of household bees. In one series of experiments a number of colonies, of foraging bees only, each took over five days before they produced any emergency queen cells; whilst colonies containing household bees of the same strain did so in less than a day. Now, we have already noted that every normal colony contains a group of bees, which I have called the "link" group (Chapter 7, p. 94), the members of which have recently become field bees and not only work in the field but also do some household work, including brood-rearing. It seems possible that such "link" bees, which will normally become whole-time foragers later, still retain some of their original interest in "queen substance". When an experimental colony, consisting entirely of field bees, is made up it is reasonable to suppose that it is those of the field bees who are still "link" bees who revert to household duties in lieu of field duties as necessity dictates: and that with their renewed brood-rearing activities any tendency they may still possess to produce emergency queen cells, in the absence of an adequate quantity of the inhibitory "queen substance", increases until eventually a few are built. It can be argued that the same sort of thing sometimes happens when some factor (such as poor foraging conditions) leads to overcrowding of the brood area of the hive in summer. The "link" bees, being unable to forage for several days on end, will probably attempt to occupy themselves with household duties including brood-rearing and thus, perhaps, increase the demand for "queen substance" to such an extent that the supply becomes insufficient to meet the demand: thus queen larvae become tolerated in the nest and swarming preparations have begun. Alternatively it is possible that neither is the actual supply of "queen substance" reduced nor the demand for it increased; but that the actual quantity that is in circulation amongst the bees becomes too small to maintain the inhibition: simply because the queen and her attendants are jostled about so much in the overcrowded conditions that the attendants who would (under other conditions) obtain sufficient inhibitory substance from the queen's body are unable to do so. Probably this last theory, which postulates a breakdown in the mechanism by which "queen substance" is distributed, is the most plausible.

These are only working hypotheses; and, however probable any one of them may be, its truth has still to be demonstrated.

When a swarm eventually leaves the parent hive it usually settles on some convenient support nearby. In all but a very few exceptional cases, it later moves off to a hollow tree or some other sheltered place in which it attempts to establish itself. It has been long known that, sometimes even before the swarm has left the parent hive (perhaps three or four days before it does so), scout bees will have been seeking a suitable place for the new home. In 1951 Lindauer published some extremely interesting observations on this subject. He described how each of a number of groups of scouts from the same colony will sometimes find, independently, different, suitable, nesting sites. On returning to the swarm, at its clustering place near the parent nest, the bees of each group indicate by means of dances, which are apparently indistinguishable from the communication dances of successful foragers (p. 204) the distance and direction from the swarm of the particular nesting site they have found. The bees that dance on the surface of the swarm may indicate the positions of three or four different nesting sites. Lindauer found that as time goes on more and more of the scouts, whose bodies are sometimes dusted with brick dust or soot acquired at the nesting sites they have discovered, come to indicate by their dances one particular nesting site. Eventually, often after several hours, nearly, if not quite, all of the dances are referring to the same place. Just how this is brought about we do not yet know; but until this has occurred the swarm is most unlikely to move off to a nesting place*. Time and again Lindauer actually observed the conversion of a scout

---

*The author was able to obtain data confirming the fact that unanimity is reached from two swarms which he observed during the summer of 1953.

from a dance which indicated the situation of one nesting place to a dance which indicated another nesting place.

In the case of one swarm Lindauer observed that the dancing scouts had "reached agreement" on one nesting site but, before their swarm could move off and occupy it, another swarm took possession of it. Apparently the scouts of the first swarm quickly became aware of this fact and ceased dancing. The swarm stayed where it was overnight and did not move off until the scouts had found another suitable nesting site.

Surely these are some of the most astonishing things that have yet been discovered in the whole realm of bee behaviour? How can bees which, one supposes, possess no powers of reasoning, reach what amounts to an agreement on one of several possible nesting sites?

CHAPTER 14

THE FOOD OF THE HONEYBEE

"But honey they gather all the year: save only in those three still months, when the weather keepeth in both bees and flowers. And it is of two sorts: the one pure and liquid, - - - the other gross and solid, which we may by like reason term ambrosia. For both serve for the food of these divine creatures."
CHARLES BUTLER (1623). *The Feminine Monarchie. 2nd ed.*

So far as we know the honeybee obtains all the materials which it requires for growth and activity from the three naturally occurring substances, nectar, pollen and water. The first two of these substances, nectar and pollen, are, of course, obtained by the foragers of a colony from the flowers of many kinds of plants. The third substance, water, is collected by the bees from many different sources such as dew and raindrops on grass and other herbage, ponds and ditches, puddles on cowpats, and so on.

NECTAR

First let us consider the composition of nectar. It is essentially a solution in water of a number of sugars, together with traces of proteins, salts, acids, enzymes and aromatic substances. In most nectars three sugars preponderate, glucose, fructose and sucrose. Most nectars, but not all, contain a high proportion of this disaccharide sugar sucrose. Often as much as 40-50% of the total sugar content of a nectar consists of it. In the process of converting nectar into honey much of the sucrose is broken down into the two simpler sugars glucose and fructose by an enzyme, invertase, which is added to it by the bees. At the same time, as the average water content of nectar is about 60%, whereas that of honey is about 20%, a good deal of water has to be evaporated away. This is brought about in the following way. On returning home with a load of nectar a forager gives most of it up to one or more household bees, each of whom seeks out a quiet part of the hive and proceeds to manipulate the nectar with her mouthparts. A bee that is processing nectar in this way opens her mandibles and regurgitates a little nectar from her honey-stomach at the same time as she moves her tongue forwards and downwards from its resting position beneath her head. The tongue's tip is then slowly raised and, almost immediately afterwards, retracted again. The complete cycle of movements is repeated over and over again, the tip being raised a little further each time and a little more nectar regurgitated, until quite a large drop has appeared and lies exposed below her still partly folded tongue. The whole drop is swallowed once more and the process repeated perhaps eighty or ninety times during the course of about twenty minutes. In this way a relatively large surface area of nectar is continually being exposed to the drying influence of the hive atmosphere, and evaporation of water from it takes place quite quickly. As much as 15% of the original water is lost.

After she has manipulated the nectar for about twenty minutes the housebee deposits it in a cell—usually an empty one or one which already contains a little semi-processed nectar. Sometimes, when room is scarce and nectar is coming in fast, it is dumped temporarily in a cell containing an egg, or even a young larva, and it may be put in the cell in this way without receiving any preliminary processing. Later on, often at night, this semi-processed nectar, or unripe honey, is taken out of the cells by household bees and subjected to further processing. Eventually when its water content has become reduced enough by evaporation (a good deal of which occurs whilst it is in a cell awaiting further manipulation by the bees) it is taken and sealed away in storage cells.

Each kind of plant secretes its own particular type of nectar which differs from the nectars of all other species of plants. As nectar is the raw material from which the bees produce honey, this results in the production of honeys which are characteristic of each species of plant. It is the minor constituents of nectars, the dextrins, salts, aromatic substances, and so forth, which are chiefly responsible for the characteristic flavours and aromas of the honeys. Most honeys are, of course, prepared by the bees from mixtures of nectars from various plants, and no two honeys are quite alike, as the proportions of the nectars used varies from colony to colony.

Only traces of protein are found in most nectars and the honeys which are prepared from them. The average protein content of honey is only 0.2 %, but heather honey may contain considerably more and Pryce-Jones has found a sample of heather honey, (i.e. prepared from nectar collected from the flowers of ling, *Calluna vulgaris*) which had a protein content of 1.8 %. The relatively high protein content of heather honey is, of course, responsible for its remarkable thixotropic properties* which make its extraction so much more difficult than that of clover and other so-called "flower" honeys. It has not yet been definitely established, however, that this protein originates in the nectar itself. It is just possible that the protein comes from the hypopharyngeal glands of the worker bees and is for some reason added by them to the nectar.

The relative proportions of the sugars present vary enormously in the nectars of different kinds of plants, and the relative amounts of glucose and fructose present play an important part in determining the rate of granulation of the restultant honey. Honey which possesses a high glucose/fructose ratio tends to granulate rapidly, whereas honey with a low glucose/fructose ratio granulates very slowly indeed.

The concentration of the sugars in nectar varies tremendously. Whereas some nectars are usually very dilute, such as that of the flowers of pear trees, others are often very concentrated. In the flowers of lime trees, for example, it is not unusual in hot, dry weather to find actual sugar crystals. However, as the concentration of the nectar in any kind of flower frequently alters not only from day to day, but even from hour to hour, average

*Honey prepared from the nectar of *Calluna vulgaris* is normally a jelly. If this jelly is disturbed by vigorous stirring it temporarily assumes a more liquid state, but on being allowed to remain undisturbed soon reverts to its former condition.

concentration figures are of little value in attempts to assess the usefulness of a plant as a honey producer. Boetius has shown, for example, that the sugar content of raspberry nectar varies from about $5\frac{1}{2}\%$ to as much as 72%.

The more dilute a sample of nectar is, the greater the amount of energy the bees will have to expend in order to concentrate it and turn it into honey. It has been calculated that unless the sugar concentration of a nectar is more than 30% any bees which collect it will expend at least as much energy in the process of converting it into honey as the sugars contained in it will provide. In fact, they are wasting their energy in collecting it. Although I feel that experimental confirmation of this figure is required before it can be accepted, the fact remains that, although they can certainly taste very low concentrations of sugar, bees will seldom collect nectar whose sugar concentration is less than 20%, even when they themselves and their colonies are almost starving.

Nectar is produced by many flowering plants in special organs known as nectaries. These are most frequently found in the flowers but can also occur on other parts of plants, when they are often called "extra-floral" nectaries to distinguish them from the much more common "floral" nectaries. Bees and other nectar-seeking insects will visit both floral and extra-floral nectaries provided that the sugar concentration of the nectar they have produced is sufficiently high. Many people must have seen bees visiting the extra-floral nectaries on the undersides of the leaves of the common laurel (Pl. 31b, p. 178), as well as those on the green buds of cornflowers (Pl. 31a, p. 178). Others are found on field beans, peaches, etc. Little is known, however, about the physical nature of nectar secretion, the process by means of which nectar is produced on the external surface of the nectary of a flower. The secretion of nectar appears to be a specific property of the cells of the nectary. It is influenced by many factors both within the plant itself and also in its external environment; and determination of the part played by any one of these is complicated by its interaction with others.

In considering some of the variable factors within the plant which influence nectar secretion, we must remember that these internal factors are themselves to some extent influenced by factors in the external environment. For example, the rate of photosynthesis, the means by which the plant builds up carbohydrate, is influenced by such things as the quantities of water and carbon dioxide available, as well as by the temperature of the environment and the light intensity.

There must be an adequate supply of carbohydrate within a plant if it is to secrete nectar. Wykes found that when the stems of freshly cut flowers of various kinds were placed in solutions of sugar (sucrose) in water, the quantity of sugar that was secreted in the nectar increased as the concentration of sugar in the feeding solution was increased, until a solution containing about 20% of sucrose was reached.

Just as our knowledge of the influence of factors within a plant on its nectar secretion is very slender, so is our knowledge of the influence on nectar secretion of the various factors in the external environment.

As long ago as 1917 Kenoyer reported that the honey crop obtained by bees from certain kinds of plants is reduced if rain falls during the flowering period. Of course honey yield is not a very satisfactory measurement of nectar secretion as it is a well-known fact that rainfall reduces the foraging activity of bees. But, rain can undoubtedly have a direct effect upon the nectar, and may actually dilute nectar in open flowers with unprotected nectaries, and thus reduce the amount of sugar obtained by the bees in each load of nectar collected. Furthermore, since nectar is hygroscopic it will, if the atmosphere is sufficiently humid, absorb moisture from the air and thus become more dilute and less attractive and valuable to bees. Equally well, of course, the nectar of flowers with relatively unprotected nectaries, such as those of the apple and plum, can lose moisture to a dry atmosphere and thus become more concentrated. Mention has already been made of the fact that in hot, dry, weather sugar crystals can sometimes be found in the flowers of lime trees.

It is reasonable to suppose that the amount of water in the soil that is available to a plant during its flowering period would have an important affect on the volume and concentration of the nectar secreted. The data available are somewhat conflicting, but there is evidence that some plants will even go on secreting nectar when they are actually wilting.

Adequate rainfall before the flowering period is almost certainly important as it is necessary for the healthy growth of the plant. But, as suggested by McLachlan, too much rain can cause too vigorous vegetative growth, and exhaustion of the carbohydrate reserves of the plant which would otherwise have been used subsequently in nectar secretion. It is interesting to note that in Britain the years in which the largest crops of honey have been obtained have often had relatively dry summers.

The influence of soil types on nectar secretion has often been discussed but, until many more data are available, all that can safely be said is that all kinds of soils which support the normal, healthy, growth of a particular species of plant appear to favour nectar secretion in this species.

As it influences the humidity of the atmosphere, temperature can have a direct effect upon the volume of the nectar that has actually been secreted by any open type of flower, in which the nectar is more or less freely exposed. Conflicting experimental results concerning the influence of temperature on the actual process of nectar secretion have been obtained, however; but it is probable that a certain minimum temperature is necessary before nectar secretion can begin. The optimum temperature for nectar secretion varies with different kinds of plants, and is almost certainly much lower for some of the flowers of spring and early summer than it is for those of midsummer. For many species of plants, including the clovers which are extremely important honey plants in England, the optimum temperature for nectar secretion appears to be rather high, probably about 80-85°F.

It has been suggested that the optimum conditions for nectar secretion occur when periods of low temperature, during which sugar accumulates in the plant's tissues, and periods of high temperature, during which the accumulated sugar is secreted in the nectar, alternate with one another. Indeed the opinion that cool nights followed by warm days favours nectar secretion by white clover plants may very well be correct.

The altitude at which a plant is growing is also said to have an effect upon its nectar secretion. It has often been stated that heather (ling) growing at or about sea level produces little nectar compared with other plants of this species growing at a thousand feet or more. However, there is a particular piece of heathland in Dorset which is only a few feet above sea level but from which it is nevertheless reported that large crops of heather honey are often obtained.

It will be realised from what has been said that our knowledge about nectar secretion is still very limited. In general it appears true to say that those conditions which result in healthy plant growth favour nectar secretion, but that excessive vegetative growth can result in a reduction both in the quantity and quality of the nectar secreted.

Ultimately, as the processes involved in nectar secretion become better understood, it may become possible to forecast the likelihood of a good nectar-flow from a given species of plant. Some years ago in Australia McLachlan suggested that the quantity of starch (carbohydrate) which is present in the sapwood of eucalyptus trees before the flowering period may have a direct relationship to the quantity of nectar ultimately produced in the flowers. His suggestion was subsequently examined by Wykes who made estimations of the amount of starch present in the sapwood at regular intervals before the trees flowered. She was able to show that when little starch was present before the flowers opened, little honey was obtained; whereas when abundant starch was present, good crops of honey were taken—provided, of course, that weather conditions during the flowering period were suitable.

Finally we must remember that even though a plant may be secreting rich nectar, honeybees may not be able to get at it on account of the structure of its flowers. A good example of a plant in which the floral structure is often the limiting factor which prevents honeybees obtaining a crop of nectar is the red clover. It has been shown that in flower heads of red clover in which the length of the corolla tubes of the florets averaged 9.6 millimetres, the nectar in a corolla tube had to rise to a height of 1.7 millimetres or more before honeybees, with their relatively short tongues, could gather it.

POLLEN

Nectar is the source of the carbohydrate portion of the diet of both larval and adult bees, and also provides them with the greater part of their water. Both adult and larval honeybees can survive on a purely carbohydrate diet for some time, but are unable to grow and develop normally without protein, all of which, except perhaps for the trace present in some nectars, is derived from pollen.

Although it has been realised for a long time that pollen is required for the development of honeybee larvae, and that a colony that is denied access to pollen cannot continue to rear brood for very long, it is only relatively recently that it has been recognised that pollen is also necessary for the full development of the young, adult bee. In 1934 Haydak showed that newly emerged bees that do not have access to stores of pollen within the hive, are not only unable to elaborate satisfactorily the brood-food which forms an essential part of the diet of worker and other larvae, but also suffer from a high mortality rate. Although adult worker honeybees can continue to elaborate some sort of brood-food for a short time in the absence of pollen, they do so at the expense of their own bodies; pollen starvation of this sort has possibly sometimes even killed adult bees. Haydak found that the average increase in the nitrogen (protein) content of five-day-old worker bees over that of newly emerged bees is 64%, and this is brought about entirely by the consumption of large quantities of pollen.

Not many years ago it used to be widely believed that stores of pollen were unnecessary, and indeed undesirable, during the winter months when little or no brood rearing was believed to be going on, although it was thought that a certain small reserve of pollen stored within the hive, near the brood, was desirable during spring and summer to supply the needs of the bees during weather unfavourable for foraging. It is now known that adequate reserves of pollen within the hive are even more necessary during those times of year when bees cannot gather fresh supplies than they are during the summer months. Indeed, it has been shown that brood rearing during February, far from being uncommon, normally occurs—but is probably limited by the quantity of pollen actually available to the bees. Today every wise beekeeper tries to see that each of his colonies has a good pollen reserve (at least four B.S. brood combs) as well as a good reserve of honey in late autumn, so that it has an excellent chance of coming through the winter in a strong condition.

Various estimates have been made of the quantity of pollen annually required by a colony of average strength; and it seems to be generally agreed that between 50 and 100 lbs. are necessary. This amounts to between two and four million bee loads of pollen, and is enough to provide for the needs of perhaps 200,000 worker bees.

The chemical composition of the pollen obtained from different kinds of flowers varies enormously, and the relative nutritional values of these different kinds of pollen to bees is unknown. The total quantity of pollen collected by a colony of bees in any given district may only be a very poor measure of the value of the pollen collected in the economy of the colony concerned. Even if we only consider variations in the protein content of pollen obtained from the flowers of different plants we find (as two American workers, Todd and Bretherick, have shown) that the amount varies from about seven to almost thirty per cent. These workers, who made chemical analyses of thirty-four different kinds of pollen, also found that approximately five per cent by weight of fresh pollen consists of fats, oils and waxes, and that about twenty per cent consists of water. Salts of potassium, phosphorus, calcium, magnesium and iron also occur in pollen in small quantities.

A number of vitamins have been found to occur in various proportions in different samples of pollen; but very little is known about the part played by any of them in honeybee nutrition. Data have, however, been obtained, both in Britain and in America, which show that a protein-rich substance such as soyabean flour, (which is sometimes fed to colonies in spring to supplement the natural pollen available to them) is made very much more valuable to the bees by the addition of a very small amount of yeast (which contains vitamins essential to some other animals) as brood rearing is thereby encouraged. Clear evidence has been obtained that soyabean-flour-yeast mixtures are more valuable to bees than the widely used soyabean-flour-dried-pollen mixtures. But the feeding of yeast alone does not appear to be of any value. The addition of a trace of yeast to soyabean flour must provide something, probably a vitamin, which enables the bees to make greater use of the soyabean flour than they would otherwise be able to do. Experiments have also been carried out in which pure vitamins, instead of yeast, have been added to soyabean flour; and it has been found that the addition of riboflavin and nicotinic acid has a beneficial effect on the rate of brood rearing. Of these two vitamins riboflavin alone was ineffective, but the addition of a mixture of the two to soyabean flour resulted in a greater increase in brood rearing than did the addition of either of them separately.

Attempts to discover the kinds of pollen which are most frequently collected by worker honeybees were made with pollen traps designed to rob a certain number of the home-coming, pollen-carrying bees of their loads as they sought to enter their hive (Pl. 39, p. 206 and Pl. 40, p. 207). No method of automatic trapping so far devised can be expected to yield really accurate results, since the proportion of the loads which are taken from the bees during a given period by any trap depends to a considerable extent upon the size of the loads themselves, which vary considerably. From some species of flowers pollen-gathering bees usually collect large sized loads (e.g. from mustard), whereas much smaller loads, which the trap is inclined to miss, are often collected from other kinds of flowers, such as the clovers. Weather conditions can also affect the size of the loads collected, for they may limit the amount of pollen available in the field, as well as the length of time spent by a foraging bee on any one expedition.

Synge has shown that colonies of bees at Rothamsted collect their pollen supplies from about one hundred different kinds of plants, but that the bulk of the pollen is obtained from a relatively small number of species. During 1946 she found that fifty-four per cent of the total number of pollen loads, collected by the bees of the colonies with which she was working, came from leguminous plants, chiefly from red and white clover, fifteen per cent from rosaceous trees and shrubs, and eleven per cent from trees such as birch, elm, beech and oak which are almost entirely wind-pollinated rather than insect-pollinated. Thus in that year these experimental colonies collected about eighty per cent of their pollen from these few sources.

The precise way in which worker honeybees collect pollen from the different kinds of flowers depends upon the structure of the flower concerned. When working such flowers as roses and dandelions the pollen-gathering bee moves about quickly amongst the stamens, often gathering them to herself with her legs, and biting at them with her mandibles, in order to dislodge the pollen grains. From some plants, such as the sweet lupin, the bees only gather the pollen fortuitously whilst seeking nectar; but they then gather the pollen together from the various parts of their bodies where it has been deposited, pack it into the pollen-baskets on their hind-legs and carry it home. When working on some kinds of flowers some of the bees, often a high proportion of them, will clean off any pollen which has become attached to their bodies and deliberately discard it. This is sometimes seen when bees are working on sunflowers: some discard the pollen, whilst others equally deliberately collect it. Hundreds of bees are often to be seen in July seeking nectar from the flowers of the common lime tree (*Tilia vulgaris*), but very few carry home any pollen, as many deliberately discard any which adheres to their bodies.

When their colonies are very short of pollen the bees will sometimes collect almost any dry, fine, powder—quite irrespective of its nutritive value. They have been known to collect flour, cement and coal dust, amongst other things. Worker honeybees gather together any pollen grains that have become attached to the many plumose hairs with which their bodies are covered and, having mixed them into a thick paste with nectar or regurgitated honey, pack them into their pollen-baskets for transport back to the hive: this was first described independently, and almost simultaneously, by a British worker, Sladen, and by an American worker, Casteel. In 1952 the whole process was re-described and beautifully illustrated by Mrs. Dorothy Hodges in a delightful little book *The Pollen Loads of the Honeybee*, in which she also illustrates, in colour as well as black-and-white, the pollen grains of many of the flowers which bees visit.

Very briefly, the process takes place as follows. When a honey-bee's body has become dusted with pollen grains she usually leaves the flower and hovers nearby, whilst she performs a regular sequence of movements by means of which the pollen grains are gathered together and placed in her pollen-baskets. Sometimes she will hang by one or both of her front-legs from the edge of a flower or other convenient support whilst she completes this operation (Pl. 33b, p. 182). Any pollen grains on the bee's face and its appendages, as well as on the first segment of her thorax, are collected together by means of brushes of stiff hairs on her front-legs and are at the same time moistened with a little re-gurgitated nectar or honey, so that they stick together. The back of her head and the other two segments of her thorax are cleaned with pollen-brushes on the middle pair of legs, which also take the pollen that has been collected together on the brushes of the bee's front-legs. The abdomen is cleaned with stiff combs on the hind-legs which are also used to take the sticky masses of pollen from the brushes on the middle pair of legs. When enough pollen has collected on the pollen combs of one hind-leg, it is raked off by the pollen rake of the other hind-leg into the pollen-press which is formed by the joint between the tibia and metatarsus. The press is then closed and the pasty mass of pollen squeezed outwards and upwards until, guided thither by guard hairs and spines, it reaches the smooth, concave, floor of the pollen-basket. In the same way the pollen on the pollen-combs of the other hind-leg is transferred into the pollen-press of its opposite number and thus to its pollen-basket. As a bee collects more and more pollen the loads in the pollen-baskets grow and are held in place by a fringe of long hairs around the sides; and very probably, as Hodges has suggested, the load on each leg is also supported on a single hair which is attached to the floor of each pollen-basket.

If pollen is very plentiful, as it often is early in the day in such

Plate 31 (opposite). EXTRA-FLORAL NECTARIES
Above.   Bee gathering nectar from extra-floral nectary on the scales of a
          cornflower bud (x 3.3).
Below.   Bee gathering nectar from extra-floral nectary on the underside of a
          laurel leaf (x 3.3).

Plate 31

flowers as poppies and dandelions, a bee may only take three or four minutes to collect a full load; but at other times she may take twenty minutes or more. This refers, of course, only to those bees that are concentrating all their efforts on pollen collection and are not dividing them between pollen- and nectar-gathering. Some flowers, such as those of the poppy and the birch tree, produce pollen but no nectar and are, therefore, only visited by bees which are deliberately seeking pollen. Many other flowers, such as the clovers, the dandelion and those of fruit trees, produce nectar as well as pollen and often a bee will collect both from the same flower on a single visit. It seems to be fairly certain that in the latter cases the bees are primarily seeking nectar and only collect and pack pollen incidentally. However, as bees that are primarily seeking pollen from flowers which produce both nectar and pollen will from time to time take a little nectar, it is not always easy to decide whether any particular bee can best be regarded as a pollen-gatherer or as a nectar-gatherer. Continued observation of the behaviour of the individual bee is the only reliable way in which such a distinction can be made.

On returning to her hive with a load of pollen the bee seeks out an empty storage cell, or one that is already partly filled with pollen (Pl. 35, p. 190), and, supporting herself on its rim, dangles her hind-legs down into it and uses her middle-legs to thrust the pollen loads off. She usually pays no further attention to the two pollen masses and leaves it to a household bee to break them up and press them down firmly into the bottom of the cell. Later a little honey may be placed on top of the pollen mass and the cell is closed with a wax capping. Pollen stored in this way keeps well and is very valuable to the colony during the colder months of the year, when no fresh pollen can be gathered. But much of the pollen brought into the hive does not get pickled away like this, but is soon used up by the household bees, who need to eat it, not only for their own full development but also to enable them to elaborate brood-food both for the larvae and adult queen.

Plate 32 (opposite). WORKER HONEYBEE COLLECTING NECTAR AND POLLEN FROM A DANDELION FLOWER (x 3.2).

WATER

Worker honeybees collect water in the field and carry it home to their hives (Pl. 34, p. 183). Many beekeepers have noticed that water collectors are about in large numbers in spring, but that the number of bees collecting water rapidly declines as soon as nectar becomes plentiful. There does, in fact, appear to be some connection between the amount of nectar available to the bees and the quantity of water they collect, at those times of year when brood-rearing is proceeding actively. There is plenty of evidence that honeybees collect water, in spring at any rate, in order to dilute honey taken from stores within the hive, to produce food for the larvae. When enough fresh nectar is available it is probable that the bees use it to feed their larvae as well as themselves, and do not draw upon their reserves of honey. The collection of water thus falls off as fresh, dilute, nectar becomes available, only to rise again whenever nectar becomes scarce and honey reserves have to be used instead. Of course it can happen that the nectar coming into the hive is so concentrated that water is required to dilute it before it is suitable for feeding to larval bees.

Bees obtain the water they require in three ways. First there is so-called "metabolic water"—that is to say, water that is produced within the bees' own bodies as a by-product of the combustion of the carbohydrates, or sugars, on which they feed. Secondly (at any rate in the case of an ill-ventilated hive) there may be water which condenses from the hive atmosphere on the inside of the walls and other cooler parts of the hive, some of it being water given off by the bees during respiration. This occurs chiefly during winter, and bees can sometimes be seen collecting this water. Finally, there are various sources of water outside the hive, in the field, and it is probably from these that the bees obtain most of the water they require during the active season.

Water is necessary for the preparation of larval food. It is also used by the honeybee for at least two other purposes. It forms an essential part of the diet of the adult worker honeybee and it is also sometimes used by bees, even in the temperate climate of Britain, for air-conditioning the hive atmosphere; for which purpose it is brought in and evaporated.

Worker honeybees are able (at all events during the winter months when little brood is present) to survive throughout long periods of confinement to the hive without access to supplies of water other than "metabolic water" and that water which may enter or condense within the hive, or is contained in the honey and pollen stores. There is no evidence that bees store water within the hive during winter. The survival of bees during winter without access to water has led many people to assume that adult worker bees do not require any more water than is available to them in the honey and pollen they eat. However, Maurizio has pointed out that when bees are kept in cages in an incubator in the laboratory and are fed on concentrated sugar syrup, honey or candy, they will often take water with avidity if it is offered to them and will survive for a longer period if water is supplied to them. It therefore seems probable that,like the larval honeybees, adult workers require water in their diet, over and above that which they obtain from honey and pollen.

CHAPTER 15

FORAGING BEHAVIOUR

"And in this great variety (of different kinds of flowers) thi⸱ is strange, that where they begin they will make an end: and not meddle with any flower of other sort, until they have their load."
CHARLES BUTLER (1623). *The Feminine Monarchie. 2nd ed.*

MANY OBSERVERS have expressed surprise at the speed with which the foragers of a colony will often become aware of new sources of nectar and pollen. Frequently bees will appear on a crop almost as soon as the first flowers open. During the summer of 1949 the author watched half-a-dozen bees collecting nectar from a small and very isolated patch of wild raspberries, growing in the middle of a thick wood, when only thirty-four of the flowers were open. How did these bees find these few flowers? Again, also during 1949, regular and frequent observations were made at a large apple tree that was flowering freely, but not until about half of the flowers had actually opened was a single bee seen collecting a load of nectar from them, although several times previously individual bees had been seen apparently investigating them. Then one day at noon two or three bees began to collect nectar from the flowers and by early afternoon the whole tree was humming with bees.

Plate 33 (opposite). POLLEN-GATHERING HONEYBEES

a (left). Bee gathering pollen from stamens of bluebell flower. Many nectar gatherers take nectar by forcing tongue between petals at base of bell instead of entering the bell itself (x 2·4).

b (right). Bee hanging from lip of bluebell flower in order to comb body and pack pollen into pollen-baskets on hind legs (x 2·4).

b

Plate 34

In order to explain such observations as these one is led to suppose that at least a proportion of the foraging force of every colony of honeybees consists of scouts, who are primarily concerned to discover new sources of food and to make their discoveries known to other members of their colony by means of the dance language (Chapter 16, p. 206). This theory was some years ago very thoroughly investigated by Oettingen-Spielberg, who carried out some very interesting experiments in a large enclosure. She used small colonies, some of which were composed entirely of bees which had emerged from their cells in an incubator and had never had any opportunities either to forage or to meet other bees who had done so, and others of which contained normal mixtures of household and foraging bees. From the results obtained she came to the conclusion that the foraging force of a colony normally contains very few scouts who seek new sources of food, and that household bees, about to begin their foraging lives, nearly all wait until they have learned, from the dances of successful foragers (including those of scout bees), the locations of the flowers from which food can be obtained and the scents associated with them. This conclusion has subsequently been confirmed by Lindauer.

Scout bees do not belong to any particular age group and it appears that a bee that has been foraging for several days will on occasion become a scout. However, once a scout bee has found a rich source of food she usually settles down to collect it. It does not seem to be a case of once a scout, always a scout.

Early in spring, and again in late summer, in many districts few crops are available for bees. The number of scouts per colony then increases considerably: many scouting bees can be seen flying hither and thither a foot or so above the ground and every now and again darting towards, hovering over, and perhaps even alighting upon, some flowers or other coloured objects which have attracted their attention. On the other hand, at those times of year when food is abundant few scouting bees can be found even close to large apiaries.

Plate 34 (opposite). HONEYBEES COLLECTING WATER FROM A LEAF-CHOKED GUTTER (x 2.1).

Another Continental worker, Opfinger, has studied and compared the learning abilities of individual honeybees and has shown how greatly they vary, thus confirming some of the conclusions reached by the author, and others when experiments were carried out with bees in mazes. In the maze experiments some bees learned the way very much more quickly than others; and a few were always very restless in the maze and showed little sign of being able to learn the correct route through it to the food which they were seeking. Opfinger also noted this restlessness of a few bees, and suggested that these few might constitute a more or less permanent scouting force, as in her experiments they seemed unable to settle down to feed at any one source for very long and were always seeking new ones. Very probably her supposition is correct, although it is clear that even such "temperamentally unstable" bees as these will (as Oettingen-Spielberg found) often settle down for a time on a crop which they have found. They would probably be useless to their colonies if they did not do so, as they would otherwise fail to recruit any bees to work upon the crops. A bee will very seldom dance on returning to her hive, and thus recruit other bees to visit a source of food which she has found, until she herself has visited it a number of times in succession.

The great majority of honeybees show little initiative in finding crops for themselves and are dependent upon information which they receive from experienced foragers. Having once found a fairly rich source of food a bee will often continue to work it until it fails, when she will usually seek an alternative source. But we can only indicate the most usual course of events, as honeybees show considerable variability in their foraging habits. To give but one example: some nectar-gatherers, perhaps most, remain quietly in the hive during those parts of the day when nectar is not available to them in the flowers of their choice; but others will learn to work another crop altogether during these intervals, returning to work the first flowers when nectar becomes available in them again.

What is it that leads scout bees to investigate new flowers as possible sources of food ? We know from the work of Lubbock, Forel, von Frisch and others, that bees can distinguish between different colours and perfumes, and can learn to associate selected colours or perfumes with food.

One sunny afternoon in spring the author was walking through a country church yard when he noticed a number of bees paying considerable attention to some small, brightly coloured, pieces of confetti which were scattered about on a path. Although the bees flew down towards and sometimes hovered over particular pieces of confetti none was seen to settle on a piece. Some purple aubretia flowers were taken and scattered singly amongst the pieces of confetti. Several scouting bees flew down to these aubretia flowers, hovered an inch or less away from them, just as they had done over various pieces of confetti, and then settled upon them, extended their tongues and searched for nectar.

In experiments which were carried out subsequently it was found that untrained, scouting, bees are attracted towards small coloured objects, particularly towards blue and yellow ones, but will very seldom alight upon them to investigate further their possibilities as sources of food unless they are also scented. It was found that, just as some colours are inherently much more attractive to scouting bees than others, so some of the scents used (all of which were extracted from flowers) were much more attractive to bees than others. Care was taken throughout these experiments to make certain that the weak floral perfumes used could not be ones with which the experimental bees had previously come into contact. Great care was also taken to see that no bee which had found food had any opportunity to carry any of it back to her colony and to recruit any of her companions to visit the source.

The data obtained in these and other similar experiments indicate that the quality that first attracts a scouting bee to investigate an artificial, or real, flower is its colour; but that on approaching the attractive object closely the bee is unlikely to investigate it further unless she is able to smell some perfume coming from it. Support for this view has been obtained quite inadvertently when marking flower-visiting bees. Sometimes the marking paint, which contains strongly smelling solvents, was placed, accidentally, too near to the head of a bee; and this caused her to behave abnormally for several minutes afterwards. When a bee that had just been carelessly marked in this way approached a flower, even one that she had frequently visited before, she hovered in front of it for a few moments and then settled on a leaf or other convenient support nearby and proceeded to clean her antennae very thoroughly. Since her antennae bear her organs of smell it was as if the bee was blowing her nose. She then approached the flower again and either entered it normally or, after hovering immediately in front of its entrance for a short time with her antennae held well forward in the characteristic attitude adopted when approaching a source of food, settled and cleaned her antennae again. Sooner or later, unless the paint had spread on to her antennae and put her olfactory organs out of action, she entered the flower again in a normal manner. It is likely that the strong smell of the solvents in the paint effectively prevented the bee from smelling the much weaker perfumes of the flowers which she approached, and that until she could smell the flowers again (after the solvents had evaporated away) the bee was very loath to enter them. Thus when the scents of groups of flowers which known bees were visiting were masked with powerful foreign odours, albeit floral ones, these bees were very reluctant to enter any of the flowers so long as the masking odours were present.

The usual part played by the scent of a flower seems to be to stimulate scouting bees to investigate them more closely, and to act as recognition marks to bees which have already learned to associate their particular scents with food. But it has been shown that really strong floral scents (such as those sometimes given off by masses of hawthorn or white clover flowers) will alone serve to attract bees from a distance to the flowers. It is almost certain that when such flowers are giving off perfume really strongly they also contain abundant nectar. On the other hand bees can find objects solely by sight stimuli; they can learn to seek odourless sugar syrup from pieces of coloured paper that are not associated with any perfume; or indeed, from flowers such as those of the bilberry which possess no scents to which bees can be trained.

These conclusions agree with those of von Frisch. Many years ago he showed that bees will remember a perfume more readily than a colour. If offered the choice of a perfume to which they have been trained, in close association with a colour unknown to them, and a colour to which they have been trained in association with a perfume unknown to them, the bees will always select the former, preferring the correct perfume and the wrong colour to the correct colour and the wrong perfume.

When a honeybee finally alights upon a flower (real or otherwise) in search of nectar, she usually explores with her tongue any narrow crevices in it. In many flowers the nectaries are situated in such places and their whereabouts are sometimes indicated by a colour which, to the human eye, contrasts with that of the general colour of the flower concerned. Over one hundred and fifty years ago the great naturalist Sprengel suggested that these could be regarded as guides to indicate to nectar-seeking visitors the position of the nectaries. This suggestion subsequently came to be regarded with a certain amount of scepticism, but we now know that the colour contrasts produced by these so-called nectar guides are often impressive to the eye of the bee and undoubtedly help her to find the nectaries. It is also clear from the results of the experiments of Huber and Lex that in the flowers of some plants the nectar guides have distinctly different odours from other parts of the flowers; whilst in others, although the scents of the nectar guides and other parts of the flowers are similar, the scents are concentrated in the nectar guides. A honeybee, with her organs of smell situated in her highly mobile antennae (and so capable of being brought very close to any surface from which a scent is emanating) is likely to be able to detect such odour differences and is thus surely aided in her search for the nectaries. That individual honeybees can detect and will follow odour trails, left either by themselves or by other bees, which are quite undetectable by man, has been demonstrated in maze and other experiments.

Ever since the time of Aristotle the constancy with which an individual honeybee will usually visit only one particular kind of flower on any single foraging expedition has frequently been noted. Darwin has pointed out that this habit is probably beneficial to her, for it would seem probable that once a bee has learned the intricacies of the structure of one particular kind of flower she will be able to collect food from flowers of this kind more expeditiously than from many other kinds of flower, each with its own peculiar structure. The results of recent work on the honeybee's ability to learn the correct route through a simple maze in order to reach a coveted source of food strongly support this idea. During her first few attempts to negotiate a maze a bee makes many mistakes, but most individuals improve very quickly and soon learn to run the maze fairly accurately.

Müller in 1882 was one of the first workers to show the constancy of the individual honeybee to one particular kind of flower, and that she restricts her foraging activity to a certain limited area in a crop of this kind of flower, sometimes for days on end. He was able to make this important discovery because he was working with bees which he had marked individually with spots of paint of different colours.

Müller's work was followed in 1906 by that of the great French naturalist Gaston Bonnier. Bonnier marked with paint all the honeybees that he found working the flowers of a long plot of buckwheat some five or six yards wide, in a large field of this plant. The following day he made observations on the distribution of all the marked bees still visiting the field and found that, apart from a few individuals near the edges, all the honeybees that were visiting the strip of buckwheat on which he had marked the bees on the previous day were marked ones. He further found that none of his marked bees were working buckwheat flowers growing more than five or six yards away from this strip. Thus he demonstrated that these particular honeybees were confining their attention to a relatively small "foraging area".

Few further observations appear to have been recorded on this interesting subject (important to both beekeeper and seed-grower) until 1931 when a Dutchman, Minderhoud, described some of his experiments. These were similar to those of Bonnier, and Minderhoud studied the movements of mass marked*bees that were working large areas covered with dandelion and white clover flowers, and found that his marked bees continually returned to flowers within a radius of not more than ten yards from

*The term "mass" marking is used in contradistinction to "individual" marking to distinguish between groups, often large groups, of bees all of which bear similar marks so that individual bees can seldom be distinguished, and bees so marked that each individual can be distinguished from any other individual.

the points at which they were originally marked whilst working flowers. Minderhoud also studied the movements of bees working on various cultivated plants, and came to the important conclusion that worker honeybees return time after time, often day after day, to an area of the crop a few yards square. In the case of a flowering crop that is covering the ground fairly thickly—i.e. with many open flowers growing close to one another—this area is probably not more than ten yards each way.

In 1936 Buzzard published the results of observations with mass marked bees in which he had found that the same bees returned continually to the flowers of the particular flowering bush upon which he had originally found and marked them. Each bush appeared to have its own population of bees, the members of which only trespassed on to a neighbouring bush of the same species when the branches of the two bushes actually interlaced with one another.

In 1943 Jeffree, Kalmus and I decided to investigate the foraging behaviour of honeybees still further. In order to simplify the problem as much as possible, we laid out a number of artificial flowers—dishes of sugar syrup fitted with reservoirs to maintain a constant supply of food and provided with a little colour—in a large, grass field.

On a good flying day in June 1942 two bees were seen to be visiting a dish which was 360 yards from their hive. These two bees were individually marked and their subsequent behaviour recorded. Not only did these two bees continue to visit the dish at very regular intervals throughout the day, each making indeed at least 150 return journeys, but they continued to do so on subsequent days. Each bee took up a characteristic stance and position on the dish so that they could be recognised without noting their markings.

In order to reach this dish these two, marked, bees had to fly over or near to other similar dishes, each of which contained an abundant supply of syrup of the same concentration and on many of which other bees, including members of their own colonies, were feeding. As far as could be determined, they never visited any dish other than this particular one, and other bees were observed to behave in the same way with the dishes on which they were first observed. It appears, both from the results of these experiments, and from the observations of Müller, Bonnier, Minderhoud and Buzzard, that as long as there is an abundance of rich syrup, or nectar, available at a given place honeybees that have found this source of food will confine their attention to it for many hours and even for days together.

In other experiments in our field, in which large numbers of marked bees were used, we found that, although a high degree of constancy to the dishes on which they had been marked was exhibited by the bees, there was some spreading to adjacent dishes. Similarly, in experiments with large numbers of marked bees that were working on various kinds of flowers, a certain small amount of drifting outwards from the point of marking was discernible. Since the supply and concentration of the food in the dishes remained constant, it was thought probable that such little wandering as did occur was due either to the disturbance caused by the operation of marking a bee, or possibly, to overcrowding.

Bees which have acquired definite foraging areas will, if they find that the food supplies have failed, often seek food from other dishes or flowers of the same kind, sometimes going half-a-mile or more to do so. This has been found to be so, using both artificial and real flowers. On several occasions the author has observed the behaviour of bees when the flower-crop they have been working on was cut. Eventually they found and moved to another field containing a crop of the same kind of flowers some distance away. Such behaviour can be of concern to seed-growers; as the cutting of a crop of flowers growing in a field near to a similar seed crop may lead to a sudden, large, influx of foragers and a high degree of competition between the foraging bees for the nectar that is available. This may lead to an increased degree of bee wandering and to a greatly increased risk of contamination of the seed crop with pollen from other plants of the same species, but of different varieties, growing outside.

Experiments with dishes of syrup have indicated that if the food supplies in some of the flowers of a crop become partially exhausted, the bees that have been working these flowers tend to extend the size of their foraging areas so as to include more flowers. But if the supplies become completely exhausted the bees will,

Plate. 35—POLLEN STORAGE
above. Worker bees on pollen storage cells (x 2.4).   *below.* Pollen-carrying bees
seeking out cells in which to place their pollen loads (x 2.4)

although compelled to seek food elsewhere, continue to visit the old, now exhausted, flowers from time to time and seek food in them.

It is reasonable to conclude that the honeybee does not as a rule wander at random over a crop of flowers but, rather, confines her attention to a small part of any crop, the size of this "foraging area" depending upon the conditions existing at the time. If the flowers are scattered, then the foraging area of an individual honeybee working the crop is likely to be much larger in extent than would be the case if the flowers were growing close to one another. The abundance of the nectar or pollen available in each flower also has a profound effect upon the size of the foraging area of each individual. A single flower may serve as the foraging area of one or more bees if it contains abundant food. The sunflower is a flower which sometimes provides such a rich source of nectar, and a newly opened poppy flower will often provide an abundance of pollen. The competition between the foragers on any particular crop will affect the average amount of food present in each flower, and exert an influence on the size of the foraging areas of individual bees; for the foraging areas of different individuals overlap one another considerably.

The foraging population of every colony is made up of at least four kinds of bees :—(a) Those satisfied foragers that have already found and settled down in foraging areas on one crop or another (b) Those foragers which have learned of definite sources of food through the dances of successful foragers and are, in consequence, seeking these crops and hunting for foraging areas of their own (c) Those foragers which have become dissatisfied with a crop which is failing and, although they may still be visiting it from time to time, are seeking new sources of food; and (d) the very small number of scout bees who leave the hive without any preconceived ideas about the kinds of flowers from which they are going to seek food. These scouts probably try a number of

Plate 36 (opposite). THE DANCE LANGUAGE OF THE HONEYBEE
The dancer (marked arrow) has just stopped dancing for a moment and is giving some of the nectar she has gathered to a bee that has been paying close attention to her dancing. Other bees near the dancer's tail are showing interest in her and have been following her dance (x 3.0).

different kinds of flowers experimentally and eventually settle down in foraging areas on that crop which at the time is the richest source of nectar or pollen. Jeffree, Kalmus and I found that when bees are offered a choice between dilute and concentrated sugar syrup in two sets of dishes that have been distributed at random in a field, they will (unless the degree of competition is too great) tend to congregate, in course of time, on those dishes which contain the syrup of the higher concentration. This is probably brought about in three ways. Those bees that discover more highly concentrated syrup tend to expose their scent-producing glands whilst feeding and thus attract other bees. Those feeding on the more dilute syrup less frequently expose their scent glands. The sight of a number of bees on a dish attracts further bees. Also, some bees taste (as it were experimentally) the syrup in dishes other than those to which they have become more or less loosely attached. If they happen to discover a dish which contains syrup of a higher concentration than that upon which they have recently been feeding they tend, apparently, to become dissatisfied with the former source of food and hence to change their allegiance and attach themselves to the dish containing the more concentrated syrup. This view is supported by an observation of von Frisch, who found that the concentration of syrup required to keep bees visiting his dishes varied in almost every experiment and depended, apparently, upon the concentration of syrup upon which the bees had last fed. The observations published by Ribbands in 1949 on the foraging method of the individual honeybee also strongly support this view and add a lot to our understanding of foraging behaviour.

Ribbands caught bees, and gave them individual distinguishing marks, whilst they were visiting the flowers in a specially planted bee garden at Rothamsted. The visits of some of these bees to flowers in this garden were then very carefully noted. The garden contained five kinds of flowers: *Limnanthes*, *Nemophila*, Shirley poppy, *Eschscholtzia* and Nasturtium. In one part of the garden there was a row of *Eschscholtzia* plants growing parallel to and only a few inches away from a row of *Limnanthes* plants. Both species were in flower and the plants were intermingling to a considerable extent. Yet, despite the close proximity of these two kinds of flowers to one another, most of the bees worked one or other kind only. However, two bees were seen to visit both kinds of flowers and the movements of one of them were recorded.

Ribbands stated that at first she seemed to be visiting both kinds of flowers indiscriminately. Later he noticed, however, that during the course of the morning the proportion of *Eschscholtzia* flowers visited on each foraging expedition was increasing, until by the sixth trip of the day just over sixty per cent of the flowers visited by this bee were *Eschscholtzias*. On her seventh and subsequent trips this bee visited more and more *Limnanthes* flowers and fewer and fewer *Eschscholtzias*. The next day this bee visited many *Limnanthes* flowers before visiting a single *Eschscholtzia*, but later in the day behaved as she had done on the day before. These changes in the proportion of *Limnanthes* to *Eschscholtzia* flowers visited were accompanied by changes in the time required to collect a load, and probably resulted from changes in the relative attractiveness of these two crops. As Ribbands has pointed out, these regular changes in the proportion of *Limnanthes* and *Eschscholtzia* flowers visited suggest that this bee was not foraging on these two kinds of flowers in a purely random manner but was apparently continually making a choice between the two. Her choice was based, presumably, on her ability to appreciate changes in the relative attractiveness as sources of food of these two kinds of flowers.

In working two crops at once this bee was undoubtedly exceptional, but there is little reason to suppose that her behaviour was abnormal. Ribbands claims that the continuous exercise of memory and choice which was so characteristic of the behaviour of this bee, was equally important, although sometimes less obvious, in the foraging behaviour of other bees which he studied.

The behaviour of this bee which was studied by Ribbands, is very similar to the behaviour of the bees that were visiting dishes in the experimental field set up by Jeffree, Kalmus and myself when the supply of syrup in a dish which a bee had been visiting with great regularity failed. Ribbands' hypothesis that the foraging bee is continually exercising memory and choice in her search for food will also explain the movement of the bees

that was observed from dishes containing syrup of low concentration to dishes containing syrup of higher concentration. No
doubt had Jeffree, Kalmus and I been working with syrup
of relatively low concentration in the dishes in our experimental field, instead of syrup of fairly high concentration, the
experimental bees might have changed their allegiance from one
dish to another much more frequently than was observed. Furthermore, it is possible that if the dishes had been placed more closely
together (in most experiments they were ten yards apart) this
might have encouraged some of the bees to visit dishes other than
those upon which each of them had settled down, and to taste the
contained syrup experimentally. However, even if the dishes had
been closer together and if several different concentrations of
syrup had been used, it seems improbable that much movement
of bees from dish to dish would have taken place unless the concentration of the syrup had varied considerably in the different
dishes. Provided that bees do not wander from one kind of
flower to another when collecting nectar much more than they
do when collecting pollen, this opinion receives some support
from observations made by Betts, Maurizio, Percival and others
of the number of loads of pollen brought home by bees which
clearly contained pollen grains deliberately collected from two or
more kinds of flowers. The number of such mixed loads appears
to be about one per cent, and certainly not more than three per
cent.

We can conclude, therefore, that honeybees may sometimes
leave the kind of flower from which they have been collecting
food for some time, to sample the pollen or nectar in a flower or
two of another kind. It appears that if a new source of food
proves to be more satisfying to a bee than the one upon which she
has been working, she will often desert the old source and establish a foraging area on the new one. But if, on the other hand, the
new source proves to be less satisfactory than the old source a
bee will soon return to the old. Even when bees have been working new sources of food for some time they still continue to visit
a few flowers of their old sources occasionally and may well return
to them once more. It seems likely that this type of behaviour
occurs much more frequently amongst bees that are foraging in

gardens, and other places where different kinds of flowers are growing very close to one another, than it does in the open country, here large areas are devoted to the production of a few flowering crops.

It might be supposed that the dances of bees that have just returned from successful foraging expeditions would recruit bees which had been foraging on other kinds of flowers from which food was not readily available at the moment. However, many years ago von Frisch demonstrated that this is seldom the case. The dancing of successful foragers only serves to stimulate two types of bees: those that have previously worked the crops from which the foragers have just returned and thus learned to associate the perfumes of their flowers with food: and those that are about to undertake their first foraging expeditions.

CHAPTER 16

DIRECTION FINDING AND COMMUNICATION

"Bees certainly have a language among themselves which they perfectly understand, tho' we do not, or at best very imperfectly."
JOHN THORLEY (1765). *An Enquiry into the Nature, Order, and Government of Bees*

WHEN CONSIDERING the highly developed social organisation of the honeybee colony many of the earlier observers of bee behaviour felt compelled to postulate the existence of a leader, usually a king bee, who directed all the activities of his subjects, the worker bees, both inside and outside the hive. In order that such directions might be given, and that the workers might co-operate to the best advantage, it was supposed that bees had some method of communicating with one another by sounds or other means. We now know, of course, that no such leader exists and that no orders are issued. But, thanks to the wonderful work of von Frisch, we also know that honeybees have a "language" which enables foragers who have found rich sources of food to recruit other members of their own colonies to work these sources, and to give the recruits information regarding the location of these sources. However, before discussing the nature of this "language" we must first consider briefly the ways in which honeybees find their way about in the field.

A few days before they undertake their first foraging expeditions young bees usually leave their hives, often around midday, on their so-called "orientation" or "play" flights, during which they apparently learn the positions of their hives relative to nearby, conspicuous, objects. If a number of bees, which have each made one or two "play" flights, are released a few yards away from their hive the majority of them return safely to their own hive even though similar hives are standing nearby. But if a number of bees of the same age as these but which have never flown, are taken from the same hive and released at the same place as the others, at most only an odd one or two arrive back at their own hive. The few that do get home probably only do so by accident, the rest either drift towards and try to enter any hive they come across or get lost altogether. This difference in home-finding ability can only be due to the fact that the bees of the first group had learned the way home, whereas the bees of the other group had not.

Wolf tested the home-finding abilities of groups of bees from a colony, at intervals after their hive had been moved to a new locality, and found that as the bees explored more and more of the territory around their hive, they showed a progressive improvement in home-finding ability when taken from their hive and released at various distances from it. Certainly those bees which have been flying for some time get to know the country for some distance around their hive; and are able to find their way home again if they are taken and released individually either in, or close to territory over which they have previously flown. The extent of the territory from points in which each bee can find her way home, undoubtedly varies under different conditions and may extend further in one direction than it does in another. Beekeepers generally assume that this distance is not more than three miles from the hive, but individual bees have undoubtedly flown much further than this from their hives and returned safely, although the "safe distance" is probably shorter as a general rule.

Wolf and others have shown that bees soon learn the positions of conspicuous visual landmarks in the field and use them for orientation. If a piece of blue tape, fifty yards long, is placed in a straight line along the ground between a hive of bees and a dish of syrup, the bees soon learn to follow the tape on their way to and from the dish. This can readily be demonstrated by displacing the dish end of the tape a few yards to one side of its original position after the bees have been flying over it for a day or two. The bees continue to follow the tape despite the fact that it has been displaced and there is no longer any food at its end. When the bees of a colony have been trained to follow a visual trail, such as a tape or a series of coloured cards, to a dish of syrup, the visual trail can be curved round so that the food end comes to lie nearer to the hive, the bees continue to follow the visual trail for half-an-hour or so. They then begin to fly home from the dish by the shortest route possible, ignoring the trail, but still continue to follow it on the outward journey. Within another hour, however, most of the bees have learned to fly directly between the dish and the hive both on their outward and homeward journeys. But if (as found by the author) the dish is now removed, many of the bees when searching for it start to follow the trail again.

Bees do not always find their way by means of visual land-marks; they make use of a number of different sensory impressions, as Wolf has shown. In one of his experiments he took a colony of bees and placed it in a large field in which there were no obvious visual landmarks. He then trained the bees to visit a dish of syrup just over one hundred and sixty yards from their hive. Let us for convenience in describing his experiment say that it was due north, in front, of the hive, as the exact position does not matter at all. The bees very soon learned to make the journey to and fro between hive and dish. Once they had done so he cap-tured and marked some of those that were feeding at the dish and liberated them individually at points due north (i.e. at the feeding place), east, west and south of the hive and at the same distance from it as the feeding place to which they had been trained. He measured the length of time that it took each bee to return home, and found that those released at the original feeding place due north of the hive returned in about half-a-minute; those released an equal distance away to the south of the

---

Plate 37 (*opposite*). INSTRUMENTAL INSEMINATION OF A QUEEN HONEYBEE WITH THE MACKENSEN AND ROBERTS APPARATUS (I)

a. Anaesthetised queen in holder with sting chamber opened.

b. Exposure of vaginal orifice, 1. (x 27).

c. Exposure of vaginal orifice, 2. (x 27).

d. Operator filling syringe with semen from a drone.

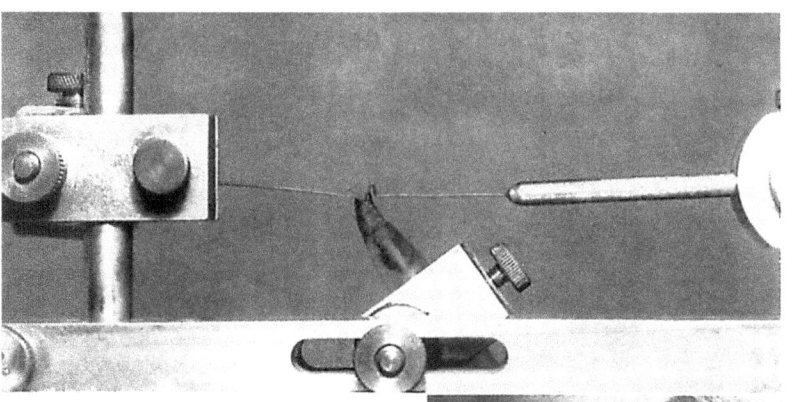

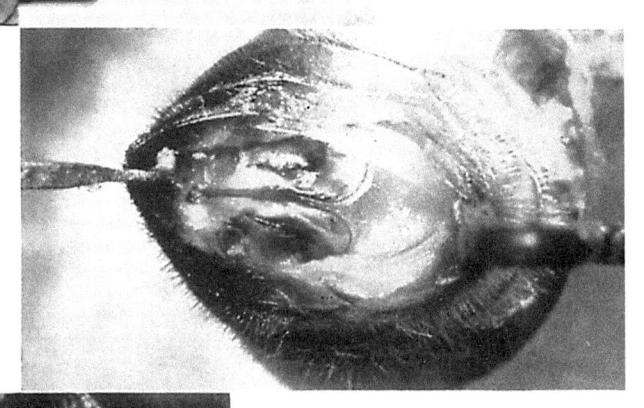

Plate 37

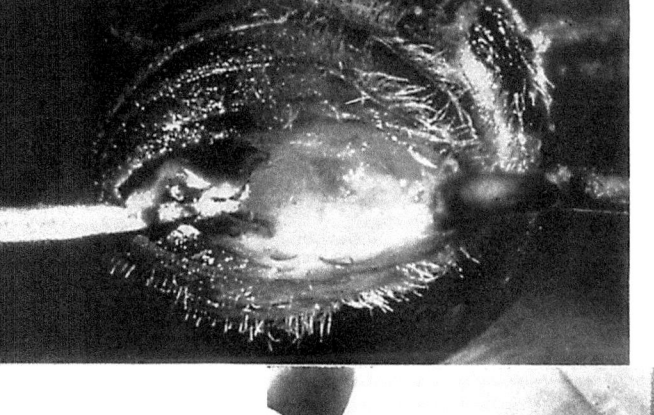

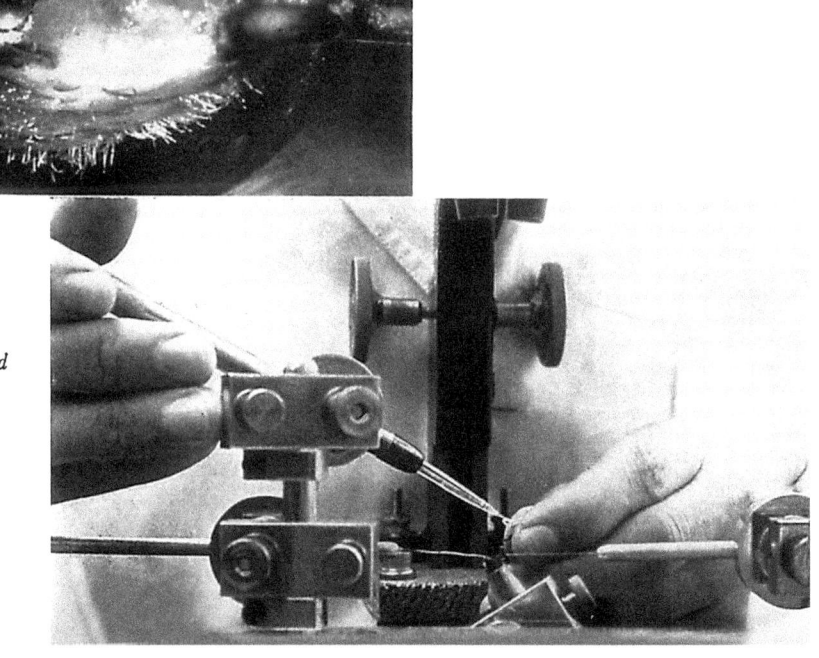

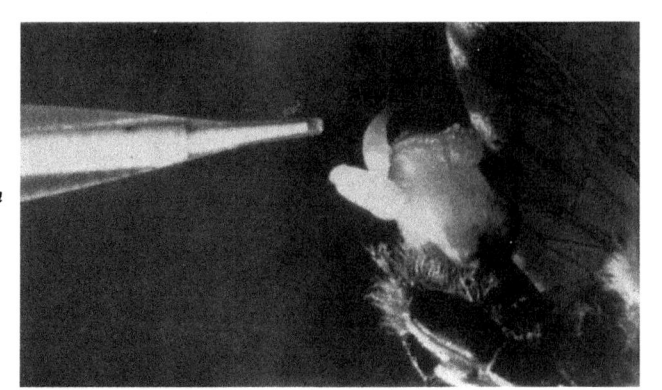

a

Plate 38

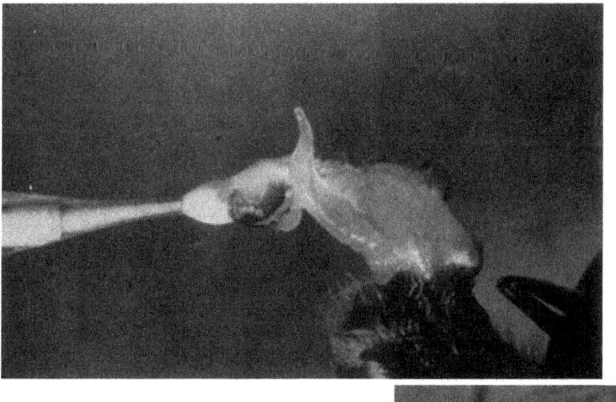

b

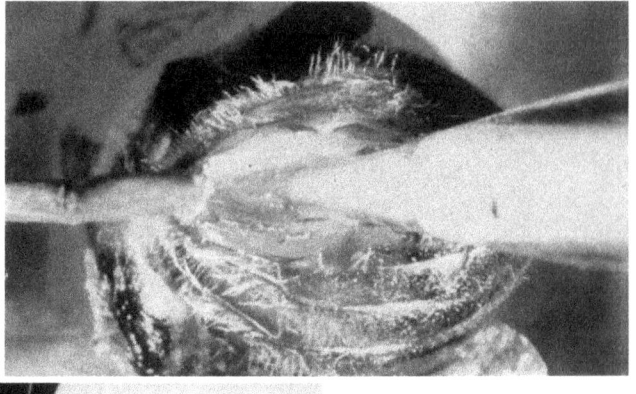

c

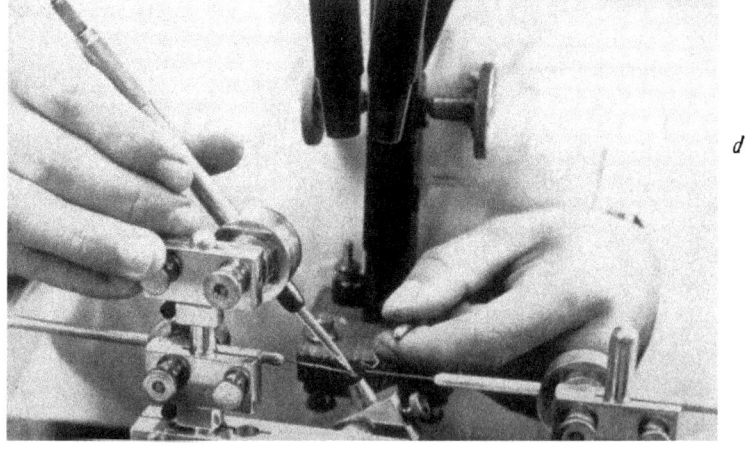

d

hive took about five times as long to get home; and those released east and west of the hive took periods of time intermediate between these two. Wolf explained these differences by suggesting that on release each bee reversed her behaviour on the way out to the dish and flew due south for a distance equal to that between hive and dish, whereupon (not finding her hive at this point) she proceeded to circle round at random until she eventually found her way home again. Since the bees that were released due south of the hive flew further south to a point twice the distance between their hive and the original feeding place and in completely the wrong direction, it naturally took them much longer to get home than bees released east and west of the hive which did not have to fly so far in the wrong direction. This suggests that the bees not only knew the direction in which they had to fly, on leaving the feeding place, in order to reach their hive, but also the "distance" or "length of time" that they had got to fly in this direction (see p. 125).

Wolf obtained further evidence by catching a number of bees at the feeding place and confining them in a cage for an hour before releasing them again at the feeding place. These bees took longer than usual to return home. During their period of confinement the sun had moved round and Wolf found that if he displaced the hive by an angle equal to that through which the sun had moved the bees returned home much more quickly. Thus he demonstrated that the bees were using the sun to help them to find their way home.

If, whilst bees are away foraging their hive is moved a short distance to one side of its original position, the returning foragers are obviously very disconcerted and will fly around and sometimes even alight and cluster at the old site. They remember the position of their hive in space. Bethe also showed, as long ago as 1898,

---

Plate 38 (opposite). INSTRUMENTAL INSEMINATION OF A QUEEN HONEYBEE WITH THE MACKENSEN AND ROBERTS APPARATUS (II)

a. Collection of semen from a drone, 1. (x 17).
b. Collection of semen from a drone, 2. (x 17).
c. Syringe tip in position in vagina of queen (x 27).
d. Operator inserting syringe tip into vagina of queen.

if bees are released from a box in open country, well away from
their hives, and their carrying-box put on one side, they will fly
around for a time, and then return to the point in space at which
they were liberated, rather than to the box.

As one would expect, the colour and (more especially) smell
of the hive have also been shown by various workers, including
von Frisch and Wolf, to help the returning bee to find her way
home. When a bee that has been "lost" eventually finds her hive
again she will often disperse scent from her scent gland and in
this way help other "lost" members of her colony to find their
way home too (Pl. 14, p. 87).

A bee learns the position and surroundings of a source of food
in the same sort of way as that in which she learns the position of
her hive relative to neighbouring objects. It has been beautifully
demonstrated by Opfinger that bees learn the colour, form
and scent of a feeding place (or flower) and of its *immediate* surroundings (14 ins. or less away) whilst they are flying towards it,
not whilst they are feeding at it nor (as many people have
supposed) whilst making their so-called "orientation flights"
around it, immediately they have left it after feeding. Opfinger
was able to show that a bee learns only the position of the feeding
place relative to more distant visual landmarks whilst making her
"orientation" flight around the food source immediately before
she returns home. She apparently learns only about the qualities
of the food itself whilst actually feeding. The foraging honeybee
appears to concentrate on one or two things at a time. During the
three or four seconds she is flying around the dish of food before
alighting at it she concentrates on memorising its colour, form,
smell and immediate surroundings. Whilst feeding at it for a
minute or two she no longer takes any notice of these characteristics but concentrates on the food itself. Then, whilst circling
around the feeding place, before going home, (for perhaps three
times as long as she did on approaching it) she concentrates
entirely upon those more distant landmarks which will help
her to find the place in space once again.

The honeybee behaves in a very similar way when hive
finding, particularly hive entrance finding, learning only the
general position of her hive in space on leaving it, and learning

details concerning its form, colour and smell when approaching it. During their "play" flights around their hives preparatory to becoming foragers young bees have, of course, the necessary opportunities to learn all of these things. At first they appear to concentrate particularly on the front of the hive itself, flying to and fro in front of and facing it, and thus learning the particular characteristics of their home so that they can find their way back again. Only later do they venture more than a yard or two away and begin to learn about the most distant surroundings of their hives.

Bees hold their antennae well forwards when approaching a hive or a flower; probably this is a sign that they are bringing their olfactory organs into action and attempting to pick up the odour associated with the objects they are approaching.

The results of some observations by von Frisch have indicated that a hitherto unsuspected sense is possessed by the worker honeybee, which helps her to find her way about in the field. He has produced evidence to show that the honeybee is able to appreciate the plane of polarization of the light that is reflected from a cloudless, blue, sky and that this ability helps her to orientate herself in space. We shall discuss this sense further when considering the way in which successful foragers are able to orientate the dances they perform on returning to their hives, and from which other members of their colonies are able to obtain information regarding the situations of particular sources of food (p. 209).

Almost all of our knowledge concerning the means by which bees obtain information about sources of food from other members of their colonies is due to the genius and to the painstaking researches of von Frisch. Although several people, such as Spitzner in 1778 and Root in 1886, had previously suggested that the well-known dancing of foragers that have just returned home serves as a means of communication to recruit other members of their colonies to collect food from favourable sources, von Frisch was the first person to demonstrate conclusively that such a "language" exists. He has also been responsible for virtually all of the subsequent experimental work that has resulted in the elucidation of at least a part of this "language".

Any discussion of the "language" of bees is apt to beguile a writer into the use of anthropomorphic phrases, which tend to suggest that the honeybee behaves deliberately in such a manner as to recruit other members of her colony to work on particular crops, about which she gives them precise directions. The author wishes to make it quite clear, therefore, that he does not know of any evidence which suggests that worker honeybees deliberately communicate with other members of their colonies. Rather, he believes, those actions on the part of bees from which other bees obtain information are performed quite unconsciously and automatically, and without any deliberate intention of passing on information.

In 1923 von Frisch published details of experiments which showed that it is by means of the dances of successful foragers on the combs in the hive that other bees are recruited to work on particular crops. He described two dances, a "round" dance and a "wag-tail" dance. In the case of the "round" dance the performer turns round in circles on the same spot, first in one direction and then in the other; in fact she traces out a figure-of-eight with its two loops more or less closely superimposed upon one another. Often a performer will continue to dance in this way on the same part of a comb for half-a-minute or more and will then, perhaps, move to another part of the comb and repeat the performance amongst a different group of bees. In the case of the "wag-tail" dance the performer also traces out a figure-of-eight, but this time the two loops are not superimposed on each other but are separated by a longer or shorter straight line. During the tracing out of this straight line between the loops the performer wags her whole body, rather than just her tail, violently from side to side. Some of the bees on the comb near a bee that is dancing in either of these ways attempt to follow her round and to touch her body with their antennae. It is by doing this that they obtain information from her (Pl. 36, p. 191).

The bees that pay close attention in this way to dancers are mostly either young bees that are just about to take up foraging for the first time, or older foragers who, perhaps on the previous day, have already learned to associate a particular kind of flower with food but who wait in the hive until they learn from the dances of other foragers, just returned from visiting them, that food is once more available in these flowers.

Von Frisch was able to demonstrate that the recruits learn the kind of perfume associated with the flowers that are yielding food. They do not learn the colours or shapes of these flowers. The recruits perceive the scents of the flowers partly directly on the bodies of the dancers with their antennae and partly by tasting some of the nectar that the dancers have collected from the flowers, which they give from time to time to bees that are following their dances. The recruits then leave the hive and seek for the perfume which they have learned from the dancers to associate with food, and are thus led to the kind of flower from which the perfume and the food were obtained.

Of the two methods by which bees learn the scent of flowers from which the dancing foragers have obtained food, the taste method has been found to be the more important, since floral scent on the bodies of foragers tends to become weak if they have to fly half-a-mile or more home. At shorter distances scent on the foragers' bodies is just as efficient as scent in the nectar which they have collected.

Von Frisch has also shown that the richer the supply of food at a given source is in quantity and quality, the more vigorously bees visiting this source dance when they get home, and the greater the number of bees they attract and recruit. If, for example, bees are allowed to feed at a dish filled with concentrated sugar syrup or honey, they will usually dance when they return home. If the syrup in the dish is now diluted with a little more water (care being taken not to add so much as to discourage the foragers from collecting it) fewer bees dance when they get home, and the dances of those that do are much less vigorous and are not continued for so long. If the syrup is diluted enough none of the bees who collect it dance. Dancing can also be reduced, or even completely inhibited, by keeping the concentration of the syrup constant but curtailing the supply so that it takes each bee much longer to collect a load. In this simple way a balance between the amount of nectar or pollen available from a particular source, and the number of bees that are recruited to work it, is achieved.

Lindauer has shown that several other factors can also affect the dancing. For instance, at those times of year when plenty of nectar is available in the field bees will seldom dance when fed on sugar syrup of low concentration, although syrup of this concentration is strong enough to cause them to dance vigorously at other times of the year when nectar is scarce. Perhaps this explains why it is that bees which have been collecting water very early, or very late, in the year, at times when little, if any, nectar is available in the field, will sometimes dance vigorously. Lindauer has also shown that bees will dance more readily when certain scents, such as the perfumes of jasmine, rose and lime flowers, are added to their syrup, than they will when the perfume of cow-parsnip is added. Indeed he obtained evidence that the addition of some perfumes, such as skatol and cow-parsnip, to syrup, not only failed to promote dancing by the bees collecting it, but actually reduced it.

In 1923 von Frisch believed that nectar-gatherers are re-cruited by the performance of "round" dances and pollen-gatherers by "wag-tail" dances, and he further believed that the only information which recruits are able to obtain from these dances concern the perfumes associated with the various sources of food. In 1946, however, he published the results of further work in which he had found that he had made a mistake in identifying the "round" dance exclusively with nectar collection and the "wag-tail" dance exclusively with pollen collection. He has now shown that the perfect "round" dance is only the first of a long series of dances which honeybees perform, the "wag-tail" dance with a long intermediate run separating the two loops of the figure-of-eight being a dance at the other end of the series. Both dances are employed, as occasion demands, to indicate the presence of either nectar or pollen.

In his 1946 and subsequent publications von Frisch has shown that, by means of this series of dances, bees that have found pro-fitable sources of either nectar or pollen in the field not only stimulate other bees to visit these sources and enable them to learn of the perfumes that are associated with them; but also enable such recruits to learn in which direction each source lies and its "distance" (p. 125) from the hive.

If a source of food is situated at a distance of approximately one hundred yards or less from her hive the dancer performs a "round" dance; if, on the other hand, it is a greater distance away she performs a "wag-tail" dance. Furthermore, as the distance between the hive and the source of food is increased the longer the straight run separating the two loops of the figure-of-eight becomes, the larger the two loops become, and the fewer become the number of figures completed in a given time. Thus, for example, if the source of food is 300 yards away from her hive a dancer will complete twenty-eight turns per minute, whereas when it is 3,000 yards away she only makes approximately nine turns per minute. This system of distance indication works at all distances between about 100 yards and three miles. Its accuracy has been shown to be enough to lead a recruit to within at least 150 yards of her objective even at the greater distances; and when the source is nearer to the hive much greater accuracy is achieved. However, when a source of food is more than three miles from the hive bees returning from it do not dance, and thus do not recruit other bees to work it.

We have already seen (p. 125) that if foragers have to fly out against a head wind their dancing becomes slower than it would if they were visiting the same feeding places on a calm day, and that a following wind has the opposite effect. It appears, therefore, that the time or energy that has to be expended in order to reach a feeding place forms the basis of the honeybee's estimation of distance. The results of some observations by Heran and Wanke support this conclusion of von Frisch. These workers noted that when bees fly uphill from their hive to a source food their dances suggest to an observer that the distance is greater than they do when the source of food lies at the same distance and at the same level as the hive. If they fly downhill to the source their dances appear to indicate that the distance is shorter. On the other hand a Russian worker, Khalifman, has obtained results which suggest that bees indicate the "distance" as determined by them on the homeward journey between food and hive. His results appear to be in direct contradiction to those of the other workers mentioned and one cannot help feeling that he has somehow misinterpreted the results of his experiments.

The system of "distance" indication helps recruits to determine how far they have got to fly in order to reach a particular source of food. It may also indicate to them how much fuel they will require to take with them for the outward journey, for it has been shown by Beutler that the amount of honey which bees carry with them when they leave their hives increases as the distance between the hive and the source of food increases. Bees that leave their hives to collect pollen, and who may not be able to collect nectar from the pollen plants visited (*e.g.*, poppies, which have no nectar), always carry much larger supplies of honey than those which are in search of nectar. It is quite possible, of course, that recruits do not learn how much honey to take with them from the dances of successful foragers, but that they have to find it out for themselves by personal experience.

From the results of some of the experiments which he described in 1946, von Frisch concluded that when a source of food is less than 25 yards from their hive the bees perform perfect "round" dances: but that as this distance is increased from 25 to about 60 yards signs of the "wag-tail" dance sometimes appear and become increasingly common as the distance is further increased, until by the time that 100 yards has been reached "round" dances have all but disappeared. When 220 yards is reached only perfect "wag-tail" dances occur. Since then Tschumi has studied the transition of the dance form from the typical, perfect, "round" dance, with the two circles of the figure-of-eight superimposed on each other, to the perfect "wag-tail" dance in which these two loops are separated by a straight run. He has described the intermediate forms as "sickle" dances on account of their form, and has found that they begin to occur when the food supply is only just over 10 yards from the hive. By the time it is 15 yards away traces of "tail wagging" sometimes creep into these "sickle" dances. It appears, therefore, that even when the food source lies close to the hive, within 100 yards, the dances of the bees visiting it can give some indication of the exact distance and do not merely indicate, as was at first supposed, that it is not more than 100 yards away.

From the "wag-tail", and to a lesser extent the "sickle", dances of successful foragers recruits obtain information not only about the distance of sources of food from the hive, but also, as von Frisch has shown, of the directions in which they lie. The direction from the hive in which a feeding place lies is shown by

Plate 39.—POLLEN TRAP (1)
above. Trap on hive. *below.* Collecting box

Plate 40.—POLLEN TRAP (II)
above. Screen pulled out. *below.* Pollen in collecting box, grid removed

the direction in which a dancer who has found it makes the straight run which separates the two loops of the figure-of-eight in her "wag-tail" dance. If the dancer makes her run vertically upwards on the comb the feeding place lies in the same direction as the sun; if she makes it downwards it lies in the opposite direction to that from which the sun is shining at the time. If the dancer runs upwards to the left at a certain angle to the vertical the feeding place lies at the same angle to the left of a straight line between the hive and the sun, and so on. It has been shown that the great majority of recruits do not deviate more than 15° from the correct direction.

We do not know how recruits judge angles so accurately, nor how they appreciate the tempo of the dance. Worker bees that have been kept out of all contact with foragers since they emerged from their cells are, on being allowed to forage, able to dance perfectly straight away and also to obtain information from the dances of other bees. So it is clear that they do not have to learn either how to dance or how to interpret a dance. However, as Lindauer has shown, it may take a young bee several days before she learns to follow a dancer's gyrations perfectly, and, until she can do so, she appears to be unable to obtain enough information to encourage her to leave the hive herself. It has been pointed out by Thorpe that the dancing of the worker honeybee is essentially an elementary form of map-making, a symbolic activity in which the direction of the force of gravity stands for the direction of the sun when a bee is dancing on the vertical face of a comb in the darkness of her hive. We have to assume that the stimulus to which the bee orientates her dance on the comb in the dark is the force of gravity, and it is surely amazing that there should be this instinctive association of the light stimulus and the gravity stimulus according to whether the bee happens to be outside or inside the hive. We must further assume that the pattern of the dance inside the hive is, so to speak, a representation of at least a part of her foraging behaviour outside the hive.

Von Frisch was himself so surprised by these discoveries of his that he tells us that he could hardly believe them and sought for an alternative explanation. For example, he examined the possibility that the bees that had found a source of food might be leaving a trail of scent between it and their hive, which trail the bees, recruited by their dances, merely followed.   However, when he sealed over the scent glands of some bees as soon as they had discovered a dish of syrup, he found that bees recruited by them nevertheless arrived safely at the dish. He could finally only conclude that bees do in fact obtain information about the distances and directions of sources of food from their hives from the dances of successful foragers.

Very occasionally bees dance on the alighting-boards of their hives or on similar horizontal surfaces *in the light*. They can also easily be caused to do so by rotation of a glass-walled observation hive, in which they are living, about its horizontal axis, or, more simply, by removing a comb on which bees are dancing from an ordinary hive and holding it in a horizontal plane. Now, when a bee dances in the light there is clearly no need for gravity to take the place of the sun, and she orientates her dance directly by the sun and makes the straight run of her "wag-tail" dance point in the actual direction of the source of food. If the horizontal surface on which a bee is dancing in this way is turned, a few degrees at a time, around its vertical axis the bee will make the necessary correction to compensate for the angle through which the surface has been turned.

Von Frisch has also been able to show that dances on a horizontal surface continue to be properly orientated as long as the dancer can see a small section of blue sky, or indeed, of sky that is lightly veiled with cloud. Should a dancer's view of the blue sky be cut off by denser cloud she may continue to dance for a time, but her dancing is disorientated and successive runs in her "wag-tail" dance point in different directions. Usually bees soon cease to dance if their views of the sky become clouded over whilst they are performing. Von Frisch's observations have also shown that when daylight is allowed to illuminate the vertical face of a comb on which bees are dancing in a glass-walled observation hive, the angle to the perpendicular is no longer indicated accurately. What is more, the degree of inaccuracy is large in the early morning and then decreases progressively as the day goes on, until by noon, when in summer the sun is more or less overhead, it has disappeared. The degree of error then increases again throughout the afternoon. If, on the other hand, the observation hive is illuminated, for purposes of observation, only with a deep red light which is invisible to honeybees (p. 127), no such errors occur. Thus it appears that it is the daylight falling on the comb that causes these errors, and that if daylight is allowed to enter a hive in this way the bees simultaneously attempt to orientate their dances partly by the sun itself and partly by gravity. Therefore the error is least at noon when the sun is directly overhead, and the sun and gravity are, so to speak, reinforcing instead of competing with one another. This can be demonstrated by throwing a beam from an electric torch on to a bee that is dancing on a perpendicular surface in the dark. If this is done the bee will either stop dancing or will re-orientate her dance so as to maintain a more or less constant angle between her straight run and the direction of the beam of light.

Von Frisch has shown that it is necessary for a bee that is dancing on a horizontal surface to be able to see the sky in order to orientate herself correctly. The bee's window to the sky need only be a few inches square, but if this small area is progressively obscured she becomes confused, more and more uncertain in her orientation, and finally stops dancing altogether. In his experiments a tube, 16 inches long and 6 inches in diameter, extended from the observation hive through the wall of a darkened hut, and it was through this that the dancing bee could see a section of the sky to the north. The feeding place was away to the west and the bees made the straight runs of their "wag-tail" dances point directly towards it. Von Frisch then reflected with a mirror an image of a part of the sky to the south down the tube so that this was all of the sky that the bees could see: the bees, who were still dancing, re-orientated their dances so that their straight runs now pointed east instead of west. From the results of this and other similar experiments it is clear that bees that cannot see the sun, and are only able to see a small piece of blue sky, are able to perceive some phenomenon which is dependent on the sun's position, and which enables them to determine the position of the sun. If their view of the sky becomes obscured by cloud they are no longer able to do this accurately, and their dances become disorientated. Von Frisch has suggested that this phenomenon is dependent on the polarization of the light that is reflected from a blue sky. Since the light that is reflected in this way is partly polarized, that is to say the light waves vibrate mostly in one plane instead of in all planes simultaneously, and the proportion of light that is polarized in this way varies in a regular way in different parts of the sky, it appears that the honeybee may thereby possess the means of knowing the position of the sun. To obtain experimental evidence that they are able to do this, von Frisch placed a "polaroid" screen over the window beneath which the bees were performing "wag-tail" dances on the horizontal surface of a comb, and then slowly rotated this analyser and noted the responses of the bees. He found that if the "polaroid" sheet was placed in such a position that the plane of polarization of the light reflected from the blue sky was not changed appreciably when it passed through it, the orientation of the dances of the bees remained unchanged and correctly indicated the direction of the food source. If, however, the "polaroid" sheet was now turned through (let us say) 10° in a clockwise or counter-clockwise direction, then the bees reorientated their dancing through the same number of degrees to the right or to the left as the case might be. Indeed the bees continued to reorientate their dancing in this way, as the "polaroid" was rotated, until it had been turned through an angle of 55° or more, whereupon their dances became as disorientated as they would have been if all the light was suddenly excluded.

Further evidence in support of this hypothesis was obtained by rearranging the bees' window so that they could see a piece of sky to the west of their hive instead of the piece to the north. When this was done it was found that once again the bees re-orientated their dancing to conform with the difference in the plane of vibration of the light reflected from these two areas of sky.

Since this remarkable discovery was made it has been shown that other widely different animals, such as ants and sandhoppers, also make use of polarized light from the sky to orientate themselves, a faculty which man finds difficult to understand as he does not possess it.

That the ways we have discussed are not the only ones in which bees can orientate themselves is shown by the fact that, by some way not yet clear, bees can still orientate their dances correctly when the sky is completely covered with cloud, or it is very foggy, and they can neither see the sun nor a piece of blue sky when they are foraging. It might be supposed that this could be explained by the bees' time-memory sense (p. 122), which might enable them to remember from what positions the sun shines at different times of the day. Von Frisch demonstrated that this is not the correct explanation by taking a colony of bees, on a day when the whole sky was covered with dense cloud, to a place where these bees had never previously been, and showing that he could train them to visit a dish for food and that some of the bees who did so performed correctly orientated dances when they returned to their hive. Even under such seemingly impossible conditions, bees still have some way of recognising the position of the sun, a way not yet fully understood. Clearly there are many problems still to be solved, and other discoveries to be made, before we understand the "language" of the bees.

Unfortunately we know very little about the evolution of the method of communication practised by the Western honeybee. During the course of some necessarily very restricted observations in Ceylon the author often observed workers of a colony of *Apis florea*, who had found a source of food, running about excitedly in all directions on the comb. Their behaviour undoubtedly excited other members of their colony and caused them to leave the nest in search of food. Workers of *Apis indica* and of *Apis mellifera* sometimes behave in a similar way when fed *within* the hive. The author has also seen workers of *Apis dorsata* behaving in a way similar to those of *Apis florea* when they had found a rich source of food, but whether or not their activity had the effect of exciting other bees and causing them to leave the nest in search of this source he was unable to determine. Workers of *A. florea* and of *A. indica* that have found a rich source of food also sometimes behave in a different way, running about in all directions on their combs waggling their bodies from side to side as they go. Is this perhaps an indication of the origin of the orientated "wag-tail" dance? Workers of *Apis indica* also perform "round" dances and orientated "wag-tail" dances which appear to be identical with those performed by *A. mellifera*. Lindauer has now

P   W.H.

shown that the top of the single comb of which an *A. florea* nest is composed is wide enough to form a horizontal dance floor on which communication dances take place. Successful *A. florea* foragers perform both "round" and "wag-tail" dances on this floor, orientating them so that the straight run separating the two loops of the figure-of-eight of the "wag-tail" dance points directly towards the source of food. In other words *A. florea* dances in the same sort of way as *A. mellifera* and *A. indica* do on those rare occasions when they dance on a horizontal surface in the light. But *A. florea*, whose comb is built in the open air (Pl. 11, p. 34), never performs orientated dances on the vertical face of the comb in the way that *A. mellifera* and *A. indica* do. It is probable, therefore, that the orientated communication dance on a horizontal surface in daylight, to perform which the dancer must be able to see a portion of unobstructed, blue sky, evolved before the orientated dance in darkness on the vertical comb face—the dance that is most frequently performed by *A. mellifera* and *A. indica* who build their multi-comb nests in hollow trees and similar dark cavities. On the other hand, the primitive *A. dorsata*, with its huge single-comb nest in the light, dances on the vertical comb face but apparently orientates its dance largely to the sun rather than to gravity.

Much of the experimental evidence for the dance language of the honeybee obtained by von Frisch and his colleagues has been checked by other research workers. Von Frisch's conclusions have, so far as I am aware, been accepted universally except by a few American workers led by Dr. A. M. Wenner who has suggested that the dances do not contain information for the bees about the distance and direction in which sources of food lie. In my opinion, however, Wenner has not produced satisfactory experimental evidence in support of his view. His chief experiment was to arrange three feeding places (1, 2 and 3) each 200 metres from a hive and 150 metres apart from each other. He then trained bees from the hive to visit feeding places 1 and 2 only, at which scented food was available, not feeding place 3, between them, where there was no food. Having trained the bees to visit feeding places 1 and 3, Wenner then provided scented food at place 2 and only unscented food at places 1 and 3, and counted the number of recruits that visited each of the three feeding places. By far the most visited the scented food at feeding place 2, leading Wenner to conclude that the scent was all important and that the recruits has obtained no information about the positions of the feeding places from the dances of the bees that had been trained to collect food at places 1 and 3. This is not, of course, the only possible explanation of the results of this and similar experiments that Wenner and his colleagues have made. Indeed, as Lindauer points out, it is quite probable that the recruits, being unable to distinguish between the dances of the bees trained to feed at places 1 and 3 because the food at them was unscented, tended to combine the information they obtained when following different dancers and so arrived at the middle feeding place. Nobody, least of all von Frisch whose pioneer work on the subject we have discussed, doubts the important part that scent, both floral and bee produced, plays in food finding. Indeed, until 1946, von Frisch himself considered the main function of the dance to be to alert potential recruits, and encourage them to leave the hive in search of particular odours. Until Wenner, or someone else, produces much more convincing evidence to the contrary— some of which should be based on experiments in which the feeding places are long distances from the hives—one must, I think, conclude that bees do indeed possess and use a "dance language" as described by von Frisch.

CHAPTER 17

EPILOGUE

A LTHOUGH man has shown a practical interest in honeybee
communities for at least four thousand years, and has learned
a good deal about them, there is still a lot that we do not know.
Furthermore this knowledge is likely to be of considerable
practical value both to beekeepers and to fruit and seed growers.
We must point to some of these gaps in our knowledge, if only to
make it clear that there is still a wide range of interesting discoveries to be made about the natural history of the honeybee.
Some aspects of it have scarcely been investigated at all. There
is no good reason why the answers to many of these problems
should not be obtained by those to whom the study of bees is a
spare time occupation rather than a profession.

Swarming is one of the most interesting and spectacular pieces
of behaviour of which the honeybee community is capable. Yet,
in spite of the fact that a full understanding of the fundamental
nature of the swarming impulse would be likely to be of very real
value to practical beekeepers, even now we still do not know what
factor (or factors) is responsible for the inception of the swarming
impulse in colonies in uncrowded hives.

Again, when we consider another very important aspect of
colony organisation—the division of labour amongst the worker
bees—we find that although we now know that it is not nearly
so elaborate or rigid as it was thought to be, and that it is readily
adaptable to the requirements of the colony, we still do not know
for certain how it is achieved. There are many similar questions
to which only tentative or incomplete answers can at present
be given.

A number of other important problems have only recently been partially solved. Only a very few years ago experimental evidence was obtained which suggests that the food that is shared so widely amongst the members of a colony plays a part in determining colony odour. It is still uncertain what parts inherited odour and absorbed odour may play in the synthesis of colony odour. Such important problems of social behaviour as colony cohesion, nest guarding, the distinction of friend from foe, and recognition of the continued presence of their queen by the workers of a colony, have only recently and partially yielded to investigation, and there is no doubt at all that there is much more still to be discovered about them.

When we turn to nutrition, either of the larvae or of the adults, the same thing holds true; and there are many other physiological problems about which we know very little indeed, such as the functions of some of the glands of the worker honeybee or of the three simple eyes, or ocelli, on top of her head.

For those who are interested in the behaviour of bees in the field there are many investigations waiting to be carried out, some of which are undoubtedly very important if we wish to use bees as pollinating agents. Until recently practically nothing was known about the field behaviour of drones and queens. Although we now know that they mate in the air we still do not know how far each sex will fly in search of a mate, nor why drones aggregate in definite places in some countries and apparently not at all in others.

A good deal is now known about the foraging behaviour of worker honeybees and of the manner in which they find their way about in the field. The well-known, epoch-making discoveries of Professor von Frisch on orientation and on the dance language of the honeybee has stirred the imaginations of many people; and workers in various countries have studied and contributed to our knowledge on this subject. There are still, however, a number of questions which we cannot answer. How are bees able to indicate correctly by means of their "wag-tail" dances, the direction in which a feeding place lies with reference to the position of the sun when the sky is covered with thick clouds? There must be aspects of this subject that still await discovery whose existence we do not at present even suspect.

The more we learn about the honeybee the more we realise how much there is still to learn.

Recent work has shown that so many of the apparently very complicated pieces of behaviour carried out by communities of honeybees, and by individual bees, are in fact brought about by very simple means. If the author has succeeded in making this clear and in interesting his readers in some of the problems of bee behaviour he will have achieved his object in writing this book.

GENERAL INDEX

DOCUMENTARY APPENDIX

THE FOLLOWING BOOKS are recommended to the reader who desires to explore the natural history of the honeybee further, and also to those who wish to study the social insects in general.

BUTLER, C. G. (1949) The Honeybee. An Introduction to her Sense-Physiology and Behaviour. Oxford Univ. Press.

ELTRINGHAM, H. (1933). The Senses of Insects. London, Methuen.

FRASER, H. M. (1951). Beekeeping in Antiquity. London, Univ. of London Press, 2nd ed.

FREE, JOHN B. and BUTLER, COLIN G. (1959). Bumblebees. London, Collins, New Naturalist.

VON FRISCH, K. (1950). Bees, their Vision, Chemical Senses, and Language. Ithaca, N.Y., Cornell Univ. Press.

VON FRISCH, K. (1954) The Dancing Bees. London, Methuen.

GROUT, R. A. (1949). The Hive and the Honeybee. Hamilton, Illinois, Dadant & Sons.

HODGES, D. (1952). The Pollen Loads of the Honeybee. London, Bee Research Assoc. Ltd.

IMMS, A. D. (1931) Social Behaviour in Insects. London, Methuen.

IMMS, A. D. (1947). Insect Natural History. London, Collins, New Naturalist.

LINDAUER, M. (1961). Communication Among Social Bees. Cambridge, Mass., Harvard Press.

MICHENER, C. D. and MICHENER, M. H. (1951). American Social Insects. New York, D. Van Nostrand Co. Inc.

RIBBANDS, C. R. (1953). The Behaviour and Social Life of Honeybees. London, Bee Research Assoc. Ltd.

RICHARDS, O. W. (1953). The Social Insects. London, MacDonald.

SNODGRASS, R. E. (1956). Anatomy of the Honeybee. Ithaca, New York, Cornell Univ. Press.

WHEELER, W. M. (1928). The Social Insects. London, Kegan Paul.

WILSON, E. O. (1971). The Insect Societies. Cambridge, Mass. Belknap (Harvard Univ. Press).

INDEX OF AUTHORS CITED